Lecture Notes in Physics

Bisher erschienen/Already published

Vol. 1: J. C. Erdmann. Wärmeleitung in Kristallen, theoretische Grundlagen und fortgeschrittene experimentelle Methoden. II, 283 Seiten. 1969.

Vol. 2: K. Hepp, Théorie de la renormalisation. III, 215 pages. 1969.

Vol. 3: A. Martin, Scattering Theory: Unitarity, Analyticity and Crossing. IV, 125 pages. 1969.

Vol. 4: G. Ludwig, Deutung des Begriffs „physikalische Theorie" und axiomatische Grundlegung der Hilbertraumstruktur der Quantenmechanik durch Hauptsätze des Messens. 1970. Vergriffen.

Vol. 5: Schaaf, The Reduction of the Product of Two Irreducible Unitary Representations of the Proper Orthochronous Quantummechanical Poincare Group. IV, 120 pages. 1970.

Vol. 6: Group Representations in Mathematics and Physics. Edited by V. Bargmann. V, 340 pages. 1970.

Vol. 7: R. Balescu, J. L. Lebowitz, I. Prigogine, P. Résibois, Z. W. Salsburg, Lectures in Statistical Physics. V, 181 pages. 1971.

Vol. 8: Proceedings of the Second International Conference on Numerical Methods in Fluid Dynamics. Edited by M. Holt. 1971. Out of print.

Vol. 9: D. W. Robinson, The Thermodynamic Pressure in Quantum Statistical Mechanics. V, 115 pages. 1971.

Vol. 10: J. M. Stewart, Non-Equilibrium-Relativistic Kinetic Theory. III, 113 pages. 1971.

Vol. 11: O. Steinmann, Pertubation Expansions in Axiomatic Field Theory. III, 126 pages. 1976.

Vol. 12: Statistical Models and Turbulence. Edited by C. Van Atta and M. Rosenblatt. Reprint of the First Edition. VIII, 492 pages. 1975.

Vol. 13: M. Ryan, Hamiltonian Cosmology. VII, 169 pages. 1972.

Vol. 14: Methods of Local and Global Differential Geometry in General Relativity. Edited by D. Farnsworth, J. Fink, J. Porter, and A. Thompson. V, 188 pages.

Vol. 15: M. Fierz, Vorlesungen zur Entwicklungsgeschichte der Mechanik. V, 97 Seiten. 1972.

Vol. 16: H.-O. Georgii, Phasenübergang 1. Art bei Gittergasmodellen. IX, 167 Seiten. 1972.

Vol. 17: Strong Interaction Physics. Edited by W. Rühl and A. Vancura. V, 405 pages. 1973.

Vol. 18: Proceedings of the Third International Conference on Numerical Methods in Fluid Mechanics, Vol. I. Edited by H. Cabannes and R. Temam. VII, 186 pages. 1973.

Vol. 19: Proceedings of the Third International Conference on Numerical Methods in Fluid Mechanics, Vol. II. Edited by H. Cabannes and R. Temam. VII, 275 pages. 1973.

Vol. 20: Statistical Mechanics and Mathematical Problems. Edited by A. Lenard. VIII, 247 pages. 1973.

Vol. 21: Optimization and Stability Problems in Continuum Mechanics. Edited by P. K. C. Wang. V, 94 pages. 1973.

Vol. 22: Proceedings of the Europhysics Study Conference on Intermediate Processes in Nuclear Reactions. Edited by N. Cindro, P. Kulišic and Th. Mayer-Kuckuk. XIV, 329 pages. 1973.

Vol. 23: Nuclear Structure Physics. Proceedings 1973. Edited by U. Smilansky, I. Talmi, and H. A. Weidenmüller. XII, 296 pages. 1973.

Vol. 24: R. F. Snipes, Statistical Mechanical Theory of the Electrolytic Transport of Nonelectrolytes. V, 210 pages. 1973.

Vol. 25: Constructive Quantum Field Theory. The 1973 "Ettore Majorana" International School of Mathematical Physics. Edited by G. Velo and A. Wightman. III, 331 pages. 1973.

Vol. 26: A. Hubert, Theorie der Domänenwände in geordneten Medien. XII, 377 Seiten. 1974.

Vol. 27: R. K. Zeytounian, Notes sur les Ecoulements Rotationnels de Fluides Parfaits. XIII, 407 pages. 1974.

Vol. 28: Lectures in Statistical Physics. Edited by W. C. Schieve and J. S. Turner. V, 342 pages. 1974.

Vol. 29: Foundations of Quantum Mechanics and Ordered Linear Spaces. Advanced Study Institute, Marburg 1973. Edited by A. Hartkämper and H. Neumann. VI, 355 pages. 1974.

Vol. 30: Polarization Nuclear Physics. Proceedings 1973. Edited by D. Fick. IX, 292 pages. 1974.

Vol. 31: Transport Phenomena. Sitges International Schools of Statistical Mechanics, June 1974. Edited by G. Kirczenow and J. Marro. XIV, 517 pages. 1974.

Vol. 32: Particles, Quantum Fields and Statistical Mechanics. Proceedings 1973. Edited by M. Alexanian and A. Zepeda. V, 132 pages. 1975.

Vol. 33: Classical and Quantum Mechanical Aspects of Heavy Ion Collisions. Proceedings 1974. Edited by H. L. Harney, P. Braun-Munzinger, and C. K. Gelbke. VII, 311 pages. 1975.

Vol. 34: One-Dimensional Conductors GPS Summer School Proceedings, 1974. Edited by H. G. Schuster. VII, 371 pages. 1975.

Vol. 35: Proceedings of the Fourth International Conference on Numerical Methods in Fluid Dynamics, 1974. Edited by R. D. Richtmyer. V, 457 pages. 1975.

Vol. 36: R. Gatignol, Théorie Cinétique des Gaz à Répartition Discrète de Vitesses. II, 219 pages. 1975.

Vol. 37: Trends in Elementary Particle Theory. Proceedings 1974. Edited by H. Rollnik and K. Dietz. V, 472 pages. 1975.

Vol. 38: Dynamical Systems, Theory and Applications. Proceedings 1974. Edited by J. Moser. VI, 624 pages. 1975.

Vol. 39: International Symposium on Mathematical Problems in Theoretical Physics. Proceedings 1975. Edited by H. Araki. XII, 562 pages. 1975.

Vol. 40: Effective Interactions and Operators in Nuclei. Proceedings 1975. Edited by B. R. Barrett. XII, 339 pages. 1975.

Vol. 41: Progress in Numerical Fluid Dynamics. Proceedings 1974. Edited by H. J. Wirz. V, 471 pages. 1975.

Vol. 42: H II Regions and Related Topics. Proceedings 1975. Edited by D. Downes and T. L. Wilson. XII, 488 pages. 1975.

Vol. 43: Laser Spectroscopy. Proceedings 1975. Edited by S. Haroche, J. C. Pebay-Peyroula, T. W. Hänsch, and S. E. Harris. X, 466 pages. 1975.

Lecture Notes in Physics

Edited by J. Ehlers, München, K. Hepp, Zürich,
R. Kippenhahn, München, H. A. Weidenmüller, Heidelberg,
and J. Zittartz, Köln

68

Y. V. Venkatesh

Energy Methods in Time-Varying System Stability and Instability Analyses

Springer-Verlag
Berlin Heidelberg GmbH 1977

Author
Y. V. Venkatesh
Department of Electrical Engineering
Indian Institute of Science
Bangalore 560012/India

ISBN 978-3-540-08430-3 ISBN 978-3-540-37160-1 (eBook)
DOI 10.1007/978-3-540-37160-1

Originally published by Springer-Verlag Berlin Heidelberg New York in 1977

2153/3140-543210

PREFACE

We frequently hear the remark that theory has outdistanced practice, and that the gap between the two is everwidening. Like all generalizations, this is a half-truth. In point of fact, stability theory is an area of considerable interest, where quite the opposite is true : There are physical systems, like the parametric amplifier with a known stable and unstable behaviour which cannot as yet be established by the general stability theories of modern times due to Lyapunov - Corduneanu, Popov and Zames - Sandberg.

Stability theory and the mathematical theory of small parameter oscillations originated in the qualitative theory of differential equations as found in the classical works of Poincare and Lyapunov who were motivated by problems in celestial mechanics. The Russian school of Chetaev, Malkin and others, in the attempt to resolve aircraft stability problems of the 1930's, pioneered work in this area. During the last twenty five years or so, stability theory has become a corner-stone of modern system theory with applications ranging from economics to satellite control. A characteristic feature of the development of the theory is that varied mathematical tools are used: Theory of functions of a complex variable, linear algebra, differential and integral inequalities, transform theory and functional analysis.

A subject of considerable study in recent years has been a nonlinear time varying feedback system, known as the generalized Aizerman - Lur'e - Postnikov system, described by a vector differential equation

$$\frac{d\underline{x}}{dt} = A_0 \underline{x} - k(t)\, \underline{b}\ \varphi(\underline{c}'\underline{x}) \qquad (!)$$

where $\underline{x}$ is an n-vector (state) ; A_0 is a given n x n constant matrix ; $\underline{b}$, $\underline{c}$ are given constant n-vectors, prime denoting transpose; k(t) is a scalar function of time and $\varphi(\cdot)$ is a scalar nonlinear function, $k(t)\varphi(\cdot)$ together constituting the feedback gain and having certain general properties. Equation (!), which arises in the theoretical analysis of various types of systems, admits, even in simple cases and definitely in general, of no closed form solution. The best that can possibly be achieved is a qualitative analysis of all the solutions of (!) for $k(t)$ and $\varphi(\cdot)$ belonging to specified classes. Evidently, a basic question to be answered is : To what extent do linearization of $\varphi(\cdot)$ and freezing of $k(t)$ at any particular value predict stability / instability of (!) ? This is a generalization of the celebrated Aizerman problem (1949) for (!) with $k(t)$ replaced by a constant K :

$$\frac{d\underline{x}}{dt} = A_o x - K \underline{b}\, \varphi(\underline{c}'\underline{x}) \qquad (!!)$$

The well known work of V.M.Popov (1959) and his frequency domain algebraic criterion for the stability of (!!) changed the course of Lyapunov - based stability analysis prevailing at that time, and bred new life into the problem of stability concerning (!). Zames and Sandberg (1963-1964) used functional analysis techniques to derive stability results similar to those of Popov and applicable to a more general class of systems having the same feedback structure as (!).

The monograph is concerned with the stability and instability analyses of systems described by (!) and also of more general systems described by integral equations having the same feedback structure as (!). Now a few words about the contents :

After introducing physical systems whose study leads to equations with time varying coefficients (Chapter 1), the problems of stability and instability for such systems are formulated along with the main methods used to resolve it (Chapter 2). For equations in form (!), the method of Lyapunov-Corduneanu (Chapter 3), and for equations in integral form, the methods of Popov (Chapter 4) and Zames - Sandberg (Chapter 5) are employed to derive stability conditions. The instability counterpart of the method of Lyapunov-Corduneanu (Chapter 6) and that of Zames - Sandberg (Chapter 7) are used to derive instability conditions for equations in corresponding forms. The exercises at the end of each chapter are a supplement to the material in the text. Capital letters along with numbers in square brackets in the body of the text refer to the corresponding item in the References at the end of the monograph. Other conventions like numbering of theorems and equations are standard.

The results of the monograph, although believed to be new and more general (for the class of equations considered) than those in some recent books, do not save us from the following conclusion : The classic stability and instability boundaries for the simplest second order linear time varying differential equation, Mathieu's or Hill's equation, cannot as yet be reproduced from any of the general results. Hence the basic stability and instability problems of time varying systems are still unsolved..

The monograph is on the whole theoretical , but it has been written by the engineer, for engineers.' It could be used as an elective one-semester course on nonlinear system stability or on Applied Mathematics to graduate students. Some familiarity with matrices, differential equations and complex variable theory is assumed on the part of the reader.

The monograph began as a series of lectures to staff and graduate students at the Institut für Regelungs- und Steuerungssysteme, University of Karlsruhe, Karlsruhe, West Germany. The author is grateful to the Alexander von Humbolt Stiftung, West Germany, for the financial support, and to Professor O.Föllinger and Professor G.Siffling for many valuable discussions.

Seminars at Budapest Muszaki Egyetem, Automatizalasi Tanszek, Budapest, Hungary and the University of Manchester Institute of Science and Technology, Manchester, England contributed to the present form of the monograph. The author wishes to thank Professor F.Csáki (Budapest) and Professor H.H.Rosenbrock (Manchester) for the stimulating time spent at their Institutes. Finally, the author owes a debt of gratitude to the authorities of the Indian Institute of Science, Bangalore for the financial support during the preparation of the manuscript, and to Professor K.S.Prabhu for encouragement and interest.

CONTENTS

CHAPTER 1

SYSTEMS AND EQUATIONS WITH TIME VARYING COEFFICIENTS

1.1. Introduction. The terms stability and instability, which are used in various ways, have been mainly associated with control. In the design of controlled systems, stability, crudely stated, implies that the system does not explode in the course of its working and that small changes in inputs or initial conditions do not lead to large changes in the system behaviour. In many cases, as, for instance, in mechanical systems first considered by Maxwell [M6] and Minorski [M10], existence of self-excited oscillations, which is undesirable, is treated as a sign of instability. In a physical system, however, we may also have 'permanent perturbations' which act for the entire duration of the system operation. Such a situation calls for different definitions of stability.

The past 20-25 years have witnessed a rapid growth of interest in time varying systems brought about by the advent of new areas of research (like parametric amplifiers and satellite stabilization) as also by the need to study in greater detail the classical vibrational problems, for instance, of columns subjected to varying axial forces. Equations with time varying coefficients also arise in the problem of infinitesimal stability of the periodic solutions of nonlinear systems. In this chapter we indicate some special features of systems and equations with time varying coefficients, and give a few illustrative examples.

1.2. Special features of systems and equations with varying coefficients.

That the response of a time varying system to a given input depends on the precise time at which the input is applied is well known. See Kaplan [K4, pp 83-84]. Here are other characteristic features of time varying systems which are of some relevance in stability / instability analysis.

1.21. Solutions and qualitative behaviour.

(a) The simplest first order linear time varying coefficient system not subject to a forcing function is described by

$$\frac{dy}{dt} + k(t)\,y = 0 \quad \text{on the interval} \quad [0,\infty). \tag{1.1}$$

The stationary or equilibrium state of the system corresponds to $y = 0$. If y_0 is the initial disturbance of the system from the equilibrium state, the solution of (1.1) is

$$y(t) = y_0 \exp\left(-\int_0^t k(\tau)d\tau\right) \tag{1.2}$$

For $k(t) \equiv K$, a positive constant, it is known that the disturbance of the system will eventually disappear. The system is then said to be asymptotically stable and, in fact, exponentially stable+ (Property 1). For $k(t) \equiv K$, a negative constant, the disturbance will eventually become very large no matter how small the initial disturbance is : The system will never return to the equilibrium state once disturbed. The system is said to be unstable and is, in fact, exponentially unstable+ (Property 2) .

Property 1 holds for arbitrary $k(t)$ if $k(t) > 0$ for all t in $[0,\infty)$, and Property 2, for arbitrary $k(t)$ if $k(t) < 0$ for all t in $[0,\infty)$. For $k(t)$ assuming both positive and negative values, stability and instability of (1.1) are determined by its average behaviour : If, for some $\varepsilon > 0$ and all $T > 0$,

$$\frac{1}{T}\int_0^T k(\tau)d\tau \geqslant \varepsilon \tag{1.3}$$

or

$$\frac{1}{T}\int_0^T k(\tau)d\tau \leqslant -\varepsilon \tag{1.4}$$

+ See Chapter 2, Sec.2.3 for definitions.

then (1.1) has Property 1 or Property 2 respectively.

Note that bounds of the type $M_1 \leq k(t) \leq M_2$ for all $t \geq 0$, where M_1 and M_2 are constants with $M_1 < M_2$, would not suffice. A distinguishing feature of the system described by (1.1) is that it could be stable for $k(t)$ assuming negative values, and unstable for $k(t)$ assuming positive values, the deciding integral inequalities being (1.3) and (1.4) respectively.

1.21. (b) If the first order system is nonlinear and is described by

$$\frac{dy}{dt} + k(t)\varphi(y) = 0 \quad \text{on the interval } [0, \infty), \tag{1.5}$$

where $\varphi(0) = 0$ and $0 < (\varphi(y)/y) < \infty$ for all $y \neq 0$, the solution to (1.5) in a closed form is complicated. The equilibrium position of (1.5) is $y = 0$. Classical stability analysis+ [B4(a)] [C6] is concerned with the behaviour of solutions of (1.5) for small deviations from this equilibrium position. For larger deviations from the equilibrium position, classical results are not valid. Hence new techniques are to be devised for stability and instability analyses 'in the large'

1.21. (c) The simplest unforced second order linear equation

$$\frac{d^2y}{dt^2} + (a - b\,k(t))y = 0 \quad \text{on the interval } [0, \infty), \tag{1.6}$$

where a and b are constants and $k(t)$ is a function of time, cannot be integrated, for an arbitrary $k(t)$, in terms of quadratures and the usual functions of mathematical physics. Furthermore, even if $k(t)$ has quite a simple form, there is no guarantee that the solutions will possess a simple structure. Thus the Mathieu equation

---

+ See Chapter 2, Sec 2.4 for a brief reference.

$$\frac{d^2y}{dt^2} + (a - b \cos \omega t)y = 0 \quad \text{on the interval } [0, \infty) \qquad (1.7)$$

presents many intricacies of analysis and has a theory of its own [M7, chapter VII] — the theory of Mathieu functions. If $k(t)$ is periodic, then (1.6) is known as the Hill equation. Stability and instability boundaries for (1.7) on the a-b parameter plane are well known [M 7, pp 113-117]. It should be noted that these boundaries have been derived using the classical method of solution for (1.7) by series expansion techniques. Similar boundaries for (1.6) are due to Borg [B 15], Krein [S 9] derived from a combination of variational techniques and classical results of mathematical analysis. See also [C 2, pp 34-90], [L 5] and [M 4].

Furthermore, if a is negative and $k(t) \equiv 0$, the solutions of (1.6) are linear combinations of $(\exp (a)^{1/2}t, \ \exp (-(a)^{1/2}t))$, and hence have unbounded solutions among them. What is characteristic of (1.6) is that $k(t)$ can be chosen in such a way that all the solutions of (1.6) are bounded. In fact, one may choose $k(t) = \cos \omega t$ and b arbitrarily small in such a way that (1.6) has only bounded solutions for all t.

In contrast with the above, if a is positive and $k(t) \equiv 0$, the solutions of (1.6) are bounded, being linear combinations of $(\cos (a)^{1/2}t$, $\sin (a)^{1/2}t)$. A periodic $k(t)$ and an arbitrarily small b can be chosen in such a way that the solutions of (1.6) are unbounded.

Suppose further we consider a linear second order constant coefficient system described by

$$\frac{d^2y}{dt^2} + \alpha \frac{dy}{dt} + K y = 0 \quad \text{on the interval } [0, \infty) \qquad (1.8)$$

where α and K are constants. If all the initial condition disturbances are to vanish after a sufficiently long time, it is necessary and sufficient that α and K are positive (Routh - Hurwitz criterion). But observe that the linear time varying system described by the equation

$$\frac{d^2y}{dt^2} + \alpha\frac{dy}{dt} + (a - b\cos\omega t)y = 0 \quad \text{on the interval } [0,\infty) \qquad (1.9)$$

is unstable, for instance, for $\alpha = 0.16$, $a = 1.0064$, $b = 0.325$ and $\omega = 2$ [M 7, p 121]. Note that, in this case, the time varying gain is always positive and has a finite value for all t in $[0,\infty)$. Conversely, with $\alpha = -0.16$, one could very well choose a,b (and ω) in such a way that the solutions of the 'negatively damped' Mathieu equation (1.9) are bounded. This has been noticed a long time ago [S 13, p 10].

Contrast the above conclusions with those for the time invariant non-linear system described by

$$\frac{d^2y}{dt^2} + \alpha\frac{dy}{dt} + \varphi(y) = 0 \quad \text{on the interval } [0,\infty) \qquad (1.10)$$

with $\alpha > 0$ $\varphi(0) = 0$ and $0 < \varepsilon \leq (\varphi(y)/y) < \infty$ for all $y \neq 0$. The equation (1.10) has solutions which tend to zero as $t \to \infty$ [A 2]. For $\alpha < 0$, the equation (1.10) has solutions which tend to $\pm\infty$ as $t \to \infty$. Note that the gain $(\varphi(y)/y)$, in both the cases, could attain a large positive value.

The implication is that there are linear time varying systems for which stability / instability requirements are more stringent than for a nonlinear time invariant system of the same order and structure. Needless to say, the situation is more complex when the system is both nonlinear and time varying.

For the example of a third order linear system in which stability / instability behaviour is distinct from the one based on constancy (or 'freezing') of coefficients, see [C 7] and also [D 5] [K 1(c)] .

1.21. (d) The phase plane analysis (or the geometric technique of Poincare) has been most successful in dealing with second order time invariant nonlinear differential equations. The reason for the success of geometric methods in dealing with equations of this type is due to the fact that the uniqueness theorem for solutions of equations of the form

$$\frac{d^2y}{dt^2} + \varphi(y, \frac{dy}{dt}) = 0$$

effectively prevents two solution curves from crossing the phase plane, (dy/dt) vs y. The phase plane method is inapplicable to second order time varying differential equations because of the explicit appearance of time in the coefficients.

1.21. (e) As evident from the above discussion, the major difficulty in the study of time varying (second order) linear differential equations lies in the determination of the 'transients' or the independent solutions of the related homogeneous differential equation. No closed form solutions are known for linear second and higher order differential equations. However, there are a few classes of equations whose solutions can be obtained in closed form. Among them is the system

$$\frac{d\underline{x}}{dt} = A(t)\,\underline{x}$$

($\underline{x}$ is an n-vector, and $A(t)$ an $n \times n$ matrix of varying coefficients) where $A(t)$ commutes with the integral of $A(t)$. The state transition matrix [Z 2] is then given by

$$\Phi(t, t_o) = \exp\left(\int_{t_o}^{t} A(\tau)d\tau\right).$$

Other similar classes of systems are given in [W 7(a)] and [W 8]. Observe that even (1.6), when written in the standard matrix form,

$$\begin{bmatrix} \frac{dx_1}{dt} \\ \frac{dx_2}{dt} \end{bmatrix} = \begin{bmatrix} 0 & 1 \\ 0 & -a + b\,k(t) \end{bmatrix} \begin{bmatrix} x_1 \\ x_2 \end{bmatrix}$$

does not satisfy the above restriction on $A(t)$. Hence the special classes of integrable systems are of limited use.

It follows that new techniques must be devised to derive information concerning the nature of the solution of time varying system equation directly from the properties of the coefficients without the intervention of explicit analytic solutions. This is important especially when the coefficients are not known exactly but belong to certain broad classes of functions.

1.22. Time varying system characteristics.

(a) The application of conventional design techniques to network problems in systems operating at relatively low frequencies often leads to impractical circuits. In addition, designs based on active RC-techniques are frequently very sensitive to small changes in element values. Alternatively, a time varying network approach to the solution of a wide class of such problems appears to be promising [F 4].

Examples are also known [B 19] of time varying R, fixed C networks having a driving point impedance with inductive reactance; time varying C, fixed R networks having a driving impedance with negative real part, and nonreciprocal two ports realised with time varying resistors, capacitors or inductors.

The implication, therefore, is that when one is constrained to use only certain types of components in building a system, a considerably larger class of systems can be realized if the component values are allowed to be time varying.

1.22. (b) Any lumped linear time invariant circuit with all the elements positive (i.e., a passive circuit in network terminology [G 6]) has the following behaviour : For any initial state of the circuit and initial time t_o, any input with bounded energy will produce an output with bounded energy. However, with the elements allowed to become time varying but positive (and bounded) at all times, the circuit may become unstable [S 2(d)] .

1.3. Examples of time varying systems.

(a) Parametric Amplifiers. A simple damped resonant circuit consists of three parameters : inductance, capacitance and resistance. The behaviour of the circuit for constant parameters is known. If the capacitance is varied periodically at a frequency approximately twice the resonant frequency of the circuit as , for instance, in frequency modulation [G 2], or parametric amplification [K 5] [M 2] , the resulting response in the circuit is oscillatory. The theoretical basis of the study of such a circuit rests on (1.6), where y is related to the response of the circuit, and $k(t)$ is the periodically varying parameter corresponding to capacitance in the circuit. In order to sustain oscillations in the circuit, an unstable solution to (1.6) is sought, i.e., a solution for which $y \to \pm \infty$ as $t \to +\infty$.

In practice, parametric energy transfer or amplification is realized by driving a nonlinear reactance in the circuit by a low frequency signal source and a single higher frequency 'pumping source' [K 5] [M 2] . The nonlinear reactance used is a varactor diode. The noise behaviour of the parametric amplifier is found to be superior to that of electron beam devices.

Other systems whose theoretical analysis leads to a study of (1.6) and higher order equations are : Condenser microphone and carbon microphone [M 2] mixers, frequency converters and nonlinear frequency multipliers [G2] consisting of some linear elements for coupling and filtering, a large amplitude steady state signal source known as a pump or local oscillator and often a d.c. bias voltage. This circuit (or system) is considered stable if the operation is always within the allowable dynamic range of the nonlinear element and a periodic steady state with the same period as that of the pump is approached asymptotically. The only restriction on the nonlinear element is that its characteristic, i.e., resistance, inductance or capacitance be described by a positive continuously differentiable function of one variable within the allowable dynamic range.

The type of parametric amplifier mentioned above is narrowband and must be used with a circulator if unilateral gain is required. In order to realise widebandwidth and stable operation, travelling wave amplifiers are constructed with variable elements at discrete points on the propagating circuit. The model is an infinite transmission line periodically loaded with varactor diodes. Across each diode is a large pumping voltage which produces a time varying capacitance [B2] . Note that such an amplifier is governed more accurately by an integral equation than by (1.6). The reason is that the system has distributed parameters and hence is governed by a partial differential equation.

A physical model frequently employed to explain the behaviour of (1.6) is a pendulum [S12, pp 189-222] . For a negative, (1.6) describes the small oscillations of an inverted pendulum with a vertical force at the pivot. For a positive, (1.6) describes the small oscillations of a normal pendulum with a vertical force at the pivot. The latter is the model for a child pumping up a swing.

1.3(b). Space systems. (i) An earth satellite, in order to perform specified functions, may be required to move in such a way that a line fixed in the satellite has a nearly constant orientation. To stabilize it in this orientation, the satellite is made into a gyroscope by permitting to spin. Hence it is natural to investigate the behaviour of spinning satellites [K 3]. Stability is related to the rate of spin and to the inertial properties of the satellite.

The study of a spinning satellite leads to the damped Mathieu equation of the type (1.9). The $\cos \omega t$ term comes in as a result of linearization about a periodic motion. Actually, linearization of the original nonlinear equations of motion of the satellite yields four first order equations with periodic coefficients [B 14(b)] . In the preceding, the disturbances were considered momentarily acting on the system, and consequently do not appear, for instance, in (1.9) used to describe the linearized equation of motion. For constantly acting disturbances, the equation(s) should contain these disturbances. Stability / instability should then be defined suitably.

(ii) The motion of an inertially oriented, earth orbiting space station acted on by gravity gradient and aerodynamic torques and controlled by reaction jet thrusters (NASA Skylab sapcedraft) is described by a set of nonlinear differential equations. The set of equations of interest are the linearized equations relative to **previously** established limit cycles about the three geometric axes of the spacecraft. The equations are [F 3] :

$$\frac{d\underline{x}}{dt} = A(t)\underline{x} - B\,\underline{u}(C\,\underline{x}) \tag{1.11}$$

where $\underline{x}$ is a 6 x 1 vector ; $A(t)$ is a periodic 6 x 6 matrix with elements having known bounds ; B is 6 x 6 constant matrix with only three elements nonzero ; and $\underline{u}$ is a 6 x 1 vector. $x_1(t)$, $x_3(t)$ and $x_5(t)$ respectively describe the linearized pitch, yaw and roll motions of the spacecraft relative to the desired limit cycle motions about these axes. Without $\underline{u}$, the system is uncontrolled and unstable, i.e., for any disturbance, the solutions do not return to the equilibrium position which is origin in the state space of the system. The nonzero elements of B and C are nonnegative. The u's are bounded **in magnitude.**

(iii) For one mode of long term operation of the Skylab [H 5] , attitude control system propellant consumption is of major concern. In this mode, the dynamic behaviour of the inertial roll angle is described by a second order nonlinear time varying differential equation. For the x-axis perpendicular to the orbit-plane mode of the Skylab, the differential equation of the inertial roll angle is

$$\frac{d^2y}{dt^2} + \alpha \sin 2(y + \omega_0 t) = \beta u$$

where $\alpha \sin 2(y + \omega_0 t)$ represents gravity gradient accleration and βu denotes reaction-control system acceleration. Roll angle should exhibit a bounded behaviour while economising on propellant consumption.

(iv) Liquid fuel rockets exhibit a type of instabilty due to self-sustained longitudinal oscillations [R 3] . Since the rocket stretches and shrinks longitudinally, it behaves like a pogo-stick, which has resulted in the nickname POGO for this type of instability. For a stability / instability analysis more accurate than the one found in [R 3] , a time varying distributed parameter model is needed. This leads to the stability and instability analyses of integral equations with varying coefficients.

1.3 (c). Aircraft. (i) The aerodynamic forces on the aircraft are functions of the speed of the aircraft. If the speed of an aircraft is changing owing to acceleration or deceleration, the aerodynamic forces will change accordingly, while the inertial properties of the aircraft will remain practically the same. As a result, if we wish to calculate the disturbed motion of the aircraft from, say, horizontal flight, the fundamental differential equation will be an equation with variable coefficients. [H 7] . It has been found [M 14] that by using a time varying gain, the motion of an aircraft following a disturbance can be made nonoscillatory and exponentially converging to the desired trajectory or motion.

(ii) The decoupled longitudinal linearized perturbation equations of motion of the tandem helicopter CH-47 [C 1] may be written in the form (1.11) where $\underline{x}$ is a 4×1 vector, A and B are respectively 4×4 and 4×2 time varying matrices, and $\underline{u}$ is a 2×1 vector. The perturbation variables are : x_1 = forward velocity in body axis ; x_2 = downward velocity in body axis ; x_3 = pitching rate ; x_4 = angle between forward body axis and horizon ; u_1 = differential collective ; u = collective. By including both level flight and hover in the flight trajectory, the eigenvalues of A move over a wide range. The helicopter is openloop (i.e., without $\underline{u}$) unstable at most of the flight conditions.

1.3 (d). Other systems.

(i) In the analysis of motion of columns with pinned ends subjected to axial forces of periodic / impulsive type, the problem is to determine whether the motion of the column is stable. Here by a stable motion we mean that the transverse oscillations of the column are bounded for all $t \geq 0$ when arbitrary initial conditions are specified. If for certain values of the frequencies of the axial force, violent lateral vibrations of the column are produced, the column is said to be unstable at these frequencies. The focal point of the analysis is the set of second order linear ordinary differential equations [L 7] [T 4, p 374] [T 5, pp 158-160] :

$$\frac{d^2y_n}{dt^2} + 2a\frac{dy_n}{dt} + (\beta_n + \alpha_n k(t))y_n = 0, \quad n = 1,2,3,\cdots, \tag{1.12}$$

where a, α_n and β_n for $n = 1,2,3,\cdots$, are constants, and $a > 0$. Decision as to the stability of the column depends on character of the set of all solutions of these equations. The set of equations (1.12) is obtained by assuming a particular series form of solution to the partial differential equation of the column satisfying the specified boundary conditions (Exercise 1.7). Evidently, a more precise analysis of the column motion would lead to an integral equation with variable coefficients.

Equations similar to (1.12) arise in problems related to the dynamic behaviour of axially moving materials, i.e., string, band saw belt, chain and pipeline vibrations [E 1] [M 13] . In all these cases, parametric resonance is caused by time dependent coefficients as, for instance, in axially moving bands and strings by a time dependent tension component which is induced by a slightly eccentric pully.

(ii) The lateral motion of a simply supported viscoelastic column, subjected to an axial load consisting of a static component P_0 and a periodic component $P(t)$, is governed by a set of ordinary differential equations with periodic coefficients of the form

$$\frac{d\underline{x}}{dt} + (A^{(o)} + \varepsilon A(t))\underline{x} = 0 \tag{1.13}$$

where $\underline{x}$ is an $n \times 1$ vector ; $A^{(o)}$ is an $n \times n$ diagonal matrix ; and ε is a small parameter [S 11] . To investigate the stability of the initially straight form of the column, it is necessary to determine whether the trivial solution $\underline{x} = \underline{0}$ of (1.13) is stable or unstable. The question is resolved by determining whether the nontrivial solutions of (1.13) are bounded or unbounded with increasing time.

(iii) The analysis of the parametrically excited pin-ended elastic column, taking nonlinearities into account, leads to the set of differential equations [B 13, Chapter 3-7]

$$\frac{d^2 y_n}{dt^2} + 2\,a\,\frac{dy_n}{dt} + (\beta_n + \alpha_n k(t))y_n + \varphi(y_n, \frac{dy_n}{dt}, \frac{d^2 y_n}{dt^2}) = 0, \; n = 1,2,3,\cdots, \qquad (1.14)$$

where a, α_n, and β_n are constants and $a > 0$. A particular form of (1.14), namely,

$$\frac{d^2 y}{dt^2} + 2\,a\,\frac{dy}{dt} + \alpha(1 - 2\beta \sin 2\,\omega t + \gamma\, y^2)y = 0, \qquad (1.15)$$

where a, α, β and γ are constants and $a > 0$, has already been studied experimentally long ago by Raman [R 1] . Some anomalies in the behaviour of parametrically excited systems are explained by introduction of a nonlinearity into the linear model.

(iv) The equation governing the behaviour of a motor control system with a time varying load and coulomb friction is given as [B 12]

$$\frac{d}{dt}\,(J(t)\,\frac{dy}{dt}) + \alpha\,\frac{dy}{dt} + \beta\, y = -\,\varphi(\frac{dy}{dt})$$

and $J(t)$ is, for simplicity, set to $J_0(1 + \gamma \cos \omega\, t) > 0$ for all t . Here the time varying load is represented as a reflected time varying inertia, and y is the shaft position.

(v) The transversal movement of a particle in an alternating gradient accelerator [P 9] is approximately described by the equation

$$\frac{d^2 y}{dt^2} + \alpha\, y + \beta\, k(\omega t)y^3 = 0. \qquad (1.16)$$

The analytical solution of (1.16) with $k(\omega\, t) \equiv 1$ for all $t \geqslant 0$ is known : It is a periodic function of time. It has been shown [P 9] that under certain conditions, the solution of (1.16) exhibits the phenomenon of parametric resonance.

(vi) The equations governing the Xenon poisoning of a nuclear reactor are given by [R 6(a)]

$$\frac{dX_1}{dt} = - aX_1 + b X_2 + (\alpha - \beta X_1)k(t)$$
$$\frac{dX_2}{dt} = - c X_2 + e k(t) \tag{1.17}$$

where X_1 is the amount of Xenon and X_2 is the amount of iodine 135, and $k(t)$ is the thermal neutron flux. Here a, b, α, β, c and e are constants and $k(t)$ is a function of time. The equations can be solved in a computer and it is then important to know whether the computed solution will be stable for small perturbations. If not the results may be meaningless.

On taking small perturbations x_1 and x_2 respectively in X_1 and X_2, we obtain from (1.17) a set of two linear differential equations with a varying coefficient $k(t)$. See [R 6(a)] for an example of a heat exchanger whose analysis leads to the study of a set of differential equations with time varying coefficients.

(vii) In the perturbation theory of collapsing spherical cavities [B 11], the governing equations are

$$k(t)\frac{d^2 y_n}{dt^2} + 3\frac{dk}{dt}\frac{dy_n}{dt} - (n-1)\frac{d^2k}{dt^2}y_n = 0, \quad n = 1,2,3,\cdots$$

where $k(t)$ is the radius of the spherical bubble, and $\dot{y}_n$'s are surface harmonics of order n. In deriving this equation, y_n's are assumed to be small, and gravity and surface tension are neglected.

It is of interest to note here that (1.6) also appears in the stability analysis of nonlinear and plasma waves.

(viii) The stability of a model reference adaptive control system

[B 14(a)] is determined ba an ordinary homogeneous differential equation of the form :

$$\frac{d^n y}{dt^n} + a_{n-1} \frac{d^{n-1} y}{dt^{n-1}} + \cdots + a_1 \frac{dy}{dt} + (\alpha - \beta\, k(t))y = 0$$

where $(\alpha - \beta\, k(t)) \triangleq K \{r(t)\}$, $r(t)$ being the input signal.

(ix) The dynamics of a Bouasse - Sardi type regulator [K 6] used in the mining industry for keeping the velocity of the descending stage approximately constant after some time interval is governed by nonlinear equations. The problem is to check the existence (or nonexistence) of oscillation in a certain sense. This gives rise to the study of the stability properties of an equation of the form

$$\frac{d^2 \underline{x}}{dt^2} + A(t)\ \varphi(\underline{x}) = \underline{f}(t)$$

where $\underline{x}$, $\underline{\varphi}$ and $\underline{f}$ are $n \times 1$ vectors, and $A(t)$ is a $n \times n$ periodic matrix.

For other examples of time varying systems, see [D 7] [F 2] [H 7] [L 2] [O 1] [P 4] and [P 5] .

1.4 Conclusions. The present day areas of research involve a study of systems which are inherently time varying or give rise, in the course of their analysis, to linear and nonlinear differential / integral equations with time varying coefficients. These equations have properties different from those of equations with time invariant coefficients. Two of the properties are : (i) Even a second order linear differential equation with a single periodic coefficient cannot be solved explicitly in closed form. (ii) Only the qualitative behaviour of solutions of these equations can possibly be analysed. Stabity and instability of solutions, which are important aspects of the qualitative behaviour, are distinct from those of solutions of equations with time invariant coefficients.

In the next chapter we formulate the problems of stability and instability for a certain class of equations with a familiar feedback structure. This class of equations is basic to the study of more general classes of equations some of which were exemplified above. Definitions of stability and instability will be introduced on the basis, for instance, of a satellite in orbit (Sec.1.3(b)) : The satellite motion or its orbit is considered stable, if, by giving a small disturbance to the motion or to its orbit, the disturbed motion or its orbit remains close to the unperturbed one for all time. More specifically,

(a) if for small disturbances the effect on the motion is small, we say that the motion is stable;

(b) if for small disturbances the effect is considerable, the motion is unstable ;

(c) if for small disturbances, the effect tends to disappear, the motion is asymptotically stable ;

(d) if, regardless of the magnitude of the disturbances, the effect on the motion tends to disappear, the motion is asymptotically stable in the large' ; and

(e) if, in the last two cases the effect on the motion tends to disappear, exponentially, the motion is 'exponentially stable in the large.'

EXERCISES

1.1 . Consider a system described by

$$p(D)y + \varphi(q(D)y) = 0 \quad \text{on the interval } [0,\infty) ,$$

where $p(D) = D^n + p_{n-1} D^{n-1} + \cdots + p_0$

$$q(D) = q_m D^m + q_{m-1} D^{m-1} + \cdots + q_0$$

are constant coefficient differential operators and the nonlinear function $\varphi(\cdot)$ is assumed to be single valued. Assume a periodic solution $r(t)$ of period T to the system equation. Find the equation of the first variation about the periodic solution. Suppose the system has a forcing function $u(t)$, and $y'(t)$, $y''(t)$ are respectively the responses for two sets of initial conditions. It is desired to find out whether these solutions approach each other. How does one derive **the relevant equations** ?

1.2 . Repeat Ex.1.1 for the system described by

$$\frac{d\underline{x}}{dt} = \underline{f}(\underline{x}, t)$$

where $\underline{x}$ and $\underline{f}$ are $n \times 1$ vectors. Assume that the vector $\underline{r}(t)$ is a periodic solution of period T to the system equation, and that $\underline{f}(\underline{x}, t)$ is continuously differentiable in $\underline{x}$. What difference would it make if $\underline{r}(t)$ were just some solution of the system equation ?

1.3. Consider the equation

$$\frac{d^2 y}{dt^2} + \alpha y + \beta y^3 = \gamma \cos 3\omega t \qquad (1.18)$$

where α, β and γ are positive constants. It is known that (1.18) has an exact sub-harmonic solution $(4\gamma/\beta)^{1/3} \cos \omega t$, provided that $\omega^2 = \alpha + 3(\gamma^2 \beta/4)^{1/3}$. It is desired to ascertain whether or not this solution is stable.

Show that the variational equation is a linear differential equation with a periodic coefficient.

1.4. The solution to the equation

$$\frac{d^2y}{dt^2} + (\alpha - 2\gamma \cos 2t)y + \beta y^3 = 0$$

where constants α, β and γ are positive, is $y = \pm 2(\gamma/\beta)^{1/2} \cos t$, provided that $\alpha = (1 - 2\gamma)$. Find its variational equation.

1.5. The differential equation of an oscillator in the range $0 < \varepsilon << 1$ is

$$(d^2y/dt^2) + \varepsilon(y^2 - 1)(dy/dt) + y = 0.$$

Take the approximate solution as $y = \alpha \sin t$. Derive the perturbational equation and indicate how one can determine the magnitude of α for stability.

1.6. Write the differential equation for the motion of a lossfree pendulum of mass m and length ℓ, the moving support being subject to a periodic force $k(t)$. Find the linearized equation.

1.7. The partial differential equation for the lateral deflection $w(z,t)$ of a column of length ℓ with a forcing function $k(t)$ at the end is described by

$$\alpha(\partial^4 w/\partial z^4) + k(t)(\partial^2 w/\partial z^2) + \beta(\partial w/\partial t) + \gamma(\partial^2 w/\partial t^2) = 0 \qquad (1.19)$$

where α, β and γ are constants related to the physical properties of the column. Boundary conditions are given as $w(0,t) = w(\ell,t)$; $(\partial^2 w/\partial z^2)|_{z=0} = (\partial^2 w/\partial z^2)|_{z=\ell} = 0$. Assume a solution to (1.19) of the form

$$w(z,t) = \sum_{n=1}^{\infty} y_n(t) \sin (n\pi z/\ell)$$

where $y_n(t)$'s for $n = 1,2,3,\cdots$ are unknown. Find the equations to be satisfied by the $y_n(t)$'s. What is the relationship between the qualitative properties of $y_n(t)$ and those of solutions to (1.19) ? How can one derive an an integral equation equivalent to (1.19) with the dependence on z removed?

CHAPTER 2

FORMULATION OF STABILITY AND INSTABILITY PROBLEMS

2.1. Introduction. As explained in Chapter 1, the mathematical models of interest to us are linear and nonlinear equations with time varying coefficients. Such models arise when the original (physical) dynamical system is inherently time varying or when, in the course of the study of the dynamical system, the equations of the system are linearized (or perturbed) around, for instance, a periodic solution. We consider here two types of models : (i) Differential Equation ; and (ii) Integral equation. We avoid questions of controllability and observability [K 1(d)] by dealing with a scalar differential equation and its equivalent form of integral equation. An advantage of such a treatment is that we can more easily compare the stability and instability results derived for the two models, and arrive at an estimate of the relative efficacy of the methods used in their analyses.

Systems in practice consist of an interconnection of subsystems. In studying an interconnected system, we would like to be able to conclude its stability and instability by considering the component subsystems and the interconnection constraint. This is particularly important in the synthesis of feedback control systems, where, for instance, the subsystem in the feedback loop is to be redesigned. In this respect, the simplest system of Fig. 2.1 plays a significant role. It consists of a linear part in the forward path and a gain in the feedback path. The linear part is assumed to be time invariant, but the feedback gain can be either linear time invariant or linear time varying or nonlinear time varying. It is logical that the system of Fig.2.1 be analysed for stability and instability with the feedback gain chosen successively in the following order : Case 1. Linear time invariant ; Case 2. Linear time varying ; and Case 3. Nonlinear time varying. Case 1 is classic, and the corresponding stability and instability criteria due to Nyquist [H 9] [Z 2] are well known. Cases 2 and 3 are of interest to us. Evidently, one way of checking the extent of generality of the stability and instability results for Cases 2 and 3 is to find out whether the Nyquist stability and instability criteria (Case 1) can be derived from them. Another, perhaps more relevant, way is to apply them to the Mathieu and Hill equations, (1.7) and (1.6), and check whether the stability and instability boundaries (a - b plane) [M 7,pp 113-117] [B 15] can be reproduced.

At present something definite can be said about the stability and related properties (like forced response characteristics) of systems modelled by Fig.2.1. As regards instability, the results available in the literature have not crystallized into any definite form.

The feedback system of Fig.2.1 can be used to model a number of physical situations which give rise to linear and nonlinear equations with time varying coefficients. For instance, the damped Mathieu equation (1.9) can be represented by Fig.2.1 with $G(s) = (1/s(s + \alpha))$, and $\mathcal{N} = (a + b \cos \omega t)$, and the nonlinear Mathieu equation (1.15) by having $G(s) = (1/s(s + 2a))$, $\mathcal{N} = \alpha(1 - 2\beta \sin 2\omega t + \gamma y^2)$. A distributed parameter system or a system with time delays and time varying coefficient could be, in some cases, represented by Fig.2.1 with an irrational (or transcendental) G(s) and a time varying $\mathcal{N}$. Note that such a system cannot be described by an ordinary differential equation.

Briefly, the feedback system of Fig.2.1 and its generalized version of Fig.2.2 form the first step in the modelling of complex physical situations. The stability and instability analyses of the systems of Fig.2.1 and Fig.2.2 can be generalized to multiloop-multivariable continuous and discrete-time feedback systems. This generalization is not attempted in the present mono-

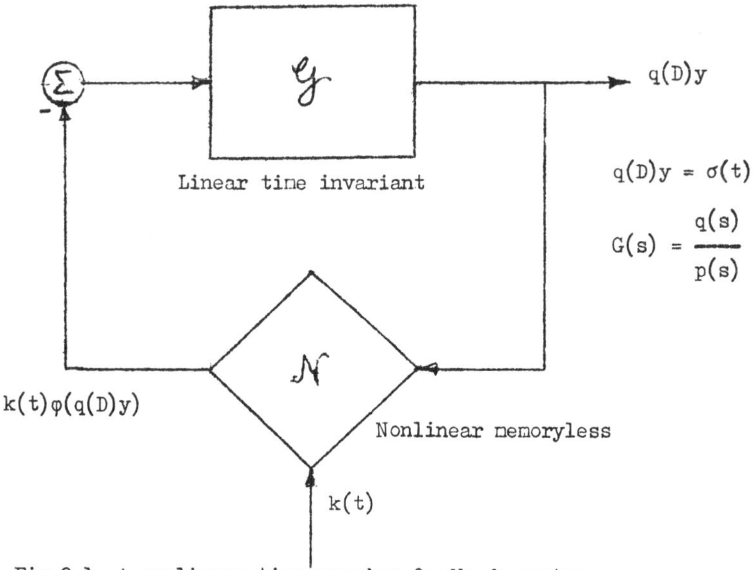

Fig.2.1. A nonlinear time varying feedback system

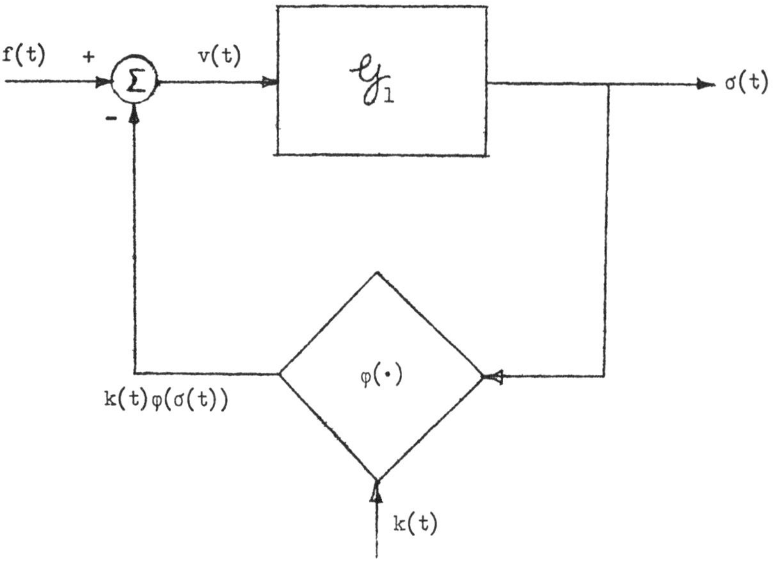

Fig.2.2. A more general nonlinear time varying feedback system

graph.

As explained in Chapter 1, an exact solution in a closed form for linear and nonlinear systems of second and higher order with time varying coefficients is unknown. That is, referring to Fig.2.1 and Fig.2.2, with complete information about $\mathcal{G}$ and $\mathcal{N}$, solution, in a closed form, to the governing equations is, in general unknown. In practice, the characteristics of $\mathcal{G}$ and $\mathcal{N}$ may change from specified values. A computer could perhaps be used to determine a numerical solution corresponding to one set of parametric values of $\mathcal{G}$ and $\mathcal{N}$, but for all possible variations of these parameters evaluation of the numerical solutions is out of the question. In these circumstances, more relevance attaches to an analysis of the qualitative behaviour of the solutions. Two important aspects of qualitative behaviour are the stability and instability properties of the solutions. As is well known, stability and instability decide the design of a system [B 16(a)].

A major premise in what follows is that $\mathcal{G}$ is known exactly but belongs to a certain specified class of functions for all t in $[0, \infty)$, the domain of definition of the system equations. Suppose, on the other hand, $\mathcal{N}$ is completely known but $\mathcal{G}$ is not, the problems of stability and instability seem to be symmetrical with those of the preceding case. Not much literature is available on the latter problems. See, however, [H 2] . Finally, incomplete knowledge about both $\mathcal{G}$ and $\mathcal{N}$ leads us nowhere, although the describing function method [H 10] for time invariant systems could be cited as dealing with this situation.

2.2 Representation of the Feedback System.

In order to get a foothold on the problems of stability and instability, we consider in sequence the following modes of mathematical representation of the basic feedback system : (a) Differential equation with zero input ;

(b) Integral equation with a restricted class of inputs ; and (c) More general integral equation with input.

2.2. (a) Differential equation with zero input.

The feedback system is represented by the scalar differential equation

$$p(D)y + k(t)\varphi(q(D)y) = 0 \quad \text{on the interval } [0, \infty), \tag{2.1}$$

where
$$p(D) = D^n + p_{n-1} D^{n-1} + \cdots + p_o , \tag{2.2}$$

$$q(D) = q_m D^m + q_{m-1} D^{m-1} + \cdots + q_o , \tag{2.3}$$

are constant coefficient differential operators with the order n of $p(D)$ at least one higher than the order m of $q(D)$, the time varying part $k(t)$ and the nonlinear part $\varphi(\cdot)$ of the feedback gain belong to certain classes defined below. The results obtained for the case $m < n$ can be transformed so as to hold for the case $m = n$ (Exercise 2.1). The case $m > n$ for the nonlinear equation (2.1) does not seem to have been studied.

Following Lur'e [L 8], it is assumed that $k(t)$ and $\varphi(\cdot)$ are not specified exactly but are assumed to belong to certain classes of functions. This is based on the premise that, in practice, the part of the system corresponding to $k(t)\varphi(\cdot)$ is the only one which may change its behaviour considerably during operation and hence may not be known exactly.

Assumption 2.21. $k(\cdot)$ is a real valued function, absolutely continuous on $[0, \infty)$ and takes values in $[\varepsilon, \infty)$ for some constant $\varepsilon > 0$. (Since $k(\cdot)$ is absolutely continuous, its derivative (dk/dt) exists almost everywhere.)

Assumption 2.22. $\varphi(\cdot)$ is a real valued function on $(-\infty, \infty)$ with the following properties : (i) $\varphi(0) = 0$; (ii) there exist constants η_1 and η_2 with $0 < \eta_1 < \eta_2$ such that $\eta_1\sigma^2 \leq \varphi(\sigma)\sigma \leq \eta_2\sigma^2$ for all $\sigma \neq 0$. This class of nonlinearities is denoted by C.

Remark 2.21. The class C of nonlinearities is very general having as its members, functions in the first and third quadrants, passing through the origin and satisfying property (ii) of Assumption 2.22. The transition from this class to the class of linear functions is abrupt. Hence it is advantageous to consider intermediate classes of nonlinearities, which include the monotone and odd monotone functions : $\varphi(\cdot) \in C_m$, the class of monotone functions if $\varphi(\cdot) \in C$ and $(\sigma_1 - \sigma_2)(\varphi(\sigma_1) - \varphi(\sigma_2)) \geq 0$ for all σ_1 and σ_2 ; $\varphi(\cdot) \in C_{mo}$, the class of odd monotone functions if $\varphi(\cdot) \in C_m$ and $\varphi(-\sigma) = -\varphi(\sigma)$ for all $\sigma \neq 0$. In fact, classes of nonlinearities, like the class of power law functions, have to be introduced in order to provide a smooth transition from class C to the class of linear functions. The transition is necessitated by our avowed intention to derive linear stability and instability criteria from the more general nonlinear stability and instability criteria.

Let $y = x_1$, $x_2 = (dx_1/dt)$, $\cdots$, $x_n = (dx_{n-1}/dt)$; and $\underline{x} = \text{col}\,[x_1, x_2, \cdots, x_n]$. Then (2.1) can be written as the vector differential equation

$$\frac{d\,\underline{x}}{d\,t} = A_o\,\underline{x} - k(t)\,\underline{b}\,\varphi(\underline{c}'\underline{x}) \tag{2.4}$$

where A_o is a matrix having the form

$$A_o = \begin{bmatrix} 0 & 1 & 0 & 0 \cdots 0 \\ 0 & 0 & 1 & 0 \cdots 0 \\ \cdot & \cdot & \cdot & \cdot \cdots \cdot \\ \cdot & \cdot & \cdot & \cdot \cdots 1 \\ -p_o & -p_1 & -p_2 & -p_3 \cdots -p_{n-1} \end{bmatrix}$$

and $\underline{b}$, $\underline{c}$ are n-vectors given by

$$\underline{b} = \text{col}\,[0, 0, \cdots, 1], \quad \underline{c} = \text{col}\,[-q_o, -q_1, \cdots, q_m, \cdots, 0].$$

Let $G(s)$ be the transfer function of the linear forward block, i.e., $G(s) = q(s)/p(s)$.

Clearly, $\underline{x}(t) \equiv \underline{0}$ is a solution of (2.4) and hence of (2.1). This solution is called the null solution (n.s.) of (2.4) and (2.1).

The following notation will be used : $\|\underline{x}\|$ denotes the norm of $\underline{x}$, where $\|\underline{x}\|^2 = \underline{x}'\underline{x}$, the prime standing for the transpose ; $\underline{x}_0$ denotes $\underline{x}(t_0)$; $\underline{x}(t; t_0, \underline{x}_0)$ denotes the solution of (2.4) which takes the value $\underline{x}_0$ for $t = t_0$. The feedback system of Fig.1 described by equation (2.1) or (2.4) with the above assumptions on $k(t)$ and $\varphi(\cdot)$ will be merely called the system (2.1) or (2.4).

2.2.(b). Integral equation with a restricted class of inputs.

The integral equation representation of the system (2.1) or (2.4) is

$$\underline{x}(t) = \exp(A_0(t - t_0))\underline{x}_0 - \int_{t_0}^{t} \exp(A(t - \tau))\underline{b}\, k(\tau)\varphi(\sigma(\tau))d\tau \tag{2.5a}$$

$$\sigma(t) \triangleq \underline{c}'\underline{x}(t) = \underline{c}'\exp(A_0(t - t_0)\underline{x}_0 - \int_{t_0}^{t} \underline{c}'\exp(A(t - \tau))\underline{b}\, k(\tau)\varphi(\sigma(\tau))d\tau \tag{2.5b}$$

Let $g(t) \triangleq \underline{c}'\exp(A_0 t)\underline{b}$ be the Laplace inverse of $G(s)$, and $f_0(t) = \underline{c}'\exp(A_0(t - t_0))\underline{x}_0$. Then (2.5b) takes the form

$$\sigma(t) = f_0(t) - \int_{t_0}^{t} g(t - \tau)k(\tau)\varphi(\sigma(\tau))d\tau \tag{2.6}$$

In this form, (2.6) can represent a class of feedback systems more general than (2.1) or (2.4). For instance, systems involving time delay and certain systems with distributed parameters may be described by (2.6).

The function $f_0(t)$ in (2.5b) is the initial condition response of the system (2.1) or (2.4). The function $f_0(t)$ in (2.6) could be treated as an input having the qualitative properties of $f_0(t)$ in (2.5b).

2.2(c). More general integral equation with input.

The feedback system of Fig. 2.2 is governed by the equations

$$v(t) = f(t) - k(t)\varphi(\sigma(t))$$
$$\sigma(t) = \sum_{i=1}^{\infty} g_i v(t - \tau_i) + \int_0^{\infty} g(\tau)v(t - \tau)d\tau \qquad (2.7)$$

for all $t \geq 0$. $k(t)$ and $\varphi(\cdot)$ satisfy Assumptions 2.1 and 2.2 respectively. g_i is a sequence with $\sum_{i=1}^{\infty} |g_i| < \infty$ and τ_i is sequence in $[0,\infty)$. Assumptions on $f(t)$ and $g(t)$ will be stated later.

Note that (2.7) includes (2.1) and (2.6) as special cases. Linear system impulse responses of the type in (2.7) occur, for instance, in circuits containing lossless transmission lines.

2.3. Preliminaries.

2.31. The starting point in the formulation of stability and instability problems for the system of Fig.1 is the time invariant linear feedback system of Fig.3 whose stability and instability properties are to be presupposed. To this end, we define stability and instability so as to make precise the intuitive explanation of Chapter 1.

Definition 2.31. The null solution of (2.1) or (2.4) is stable if for every $\varepsilon > 0$, there exists a $\delta(\varepsilon\,; t_o)$ such that if $\| \underline{x}_o \| < \delta$ there follows $\| \underline{x}(t; t_o, \underline{x}_o) \| < \varepsilon$ for $t \geq t_o \geq 0$.

Definition 2.32. The null solution of (2.1) or (2.4) is asymptotically stable if it is stable, and, in addition, there exists a $\delta_o(t_o) > 0$ with the property that if $\| \underline{x}_o \| < \delta$, then $\lim_{t\to\infty} \underline{x}(t; t_o, \underline{x}_o) = \underline{0}$.

Definition 2.33. The forced system

$$p(D)y + k(t)\varphi(q(D)y) = f(t)$$

or equivalently, $$\frac{d\underline{x}}{dt} = A_o \underline{x} - k(t)\underline{b}\,\varphi(\underline{c}'\underline{x}) + \underline{f}(t) \qquad t \in [0, \infty) \qquad (2.8)$$

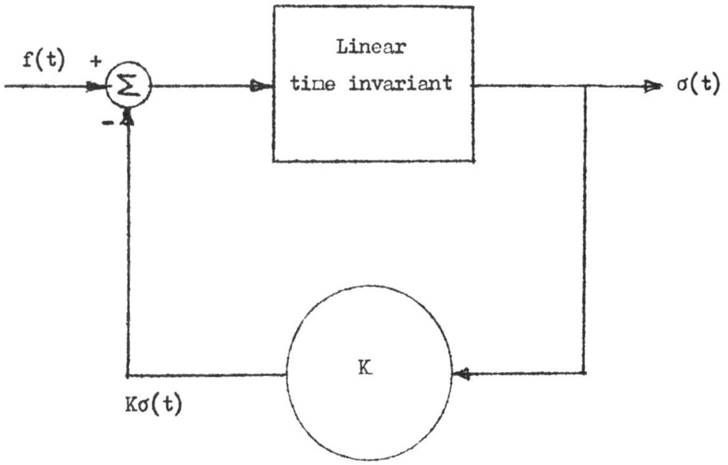

Fig.2.3(a). The basic linear time invariant feedback system

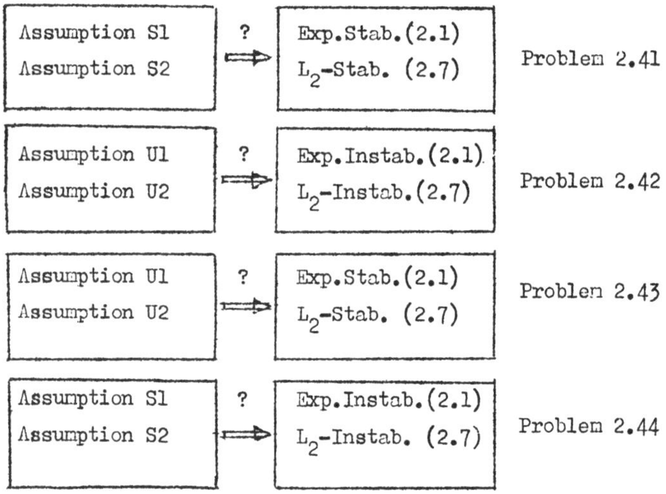

Fig.2.3(b). Fundamental stability and instability problems

where $\underline{f}(t) = \text{col}\left[0,0,\cdots,0,f(t)\right]$, is bounded-input bounded-output stable if for every $\eta_1 \geq 0$, every $\eta_2 \geq 0$, there is a number η_3 such that

$$\| \underline{x}(t;\ t_o,\ x_o) \| \leq \eta_3 \qquad \text{for all} \quad t \geq t_o \geq 0$$

for every initial condition $\underline{x}_o$ with $\| \underline{x}_o \| < \eta_1$ and every function $\underline{f}(t)$ with $\| \underline{f}(t) \| < \eta_2$.

Definition 2.34. The null solution of (2.1) or (2.4) is exponentially stable if there exist positive constants ε_1, ε_2, such that, for $t \geq t_o \geq 0$,

$$\| \underline{x}(t;\ t_o,\ \underline{x}_o) \| \leq \varepsilon_2 \, \|\underline{x}_o\| \, \exp(-\varepsilon_1(t - t_o)). \tag{2.9}$$

Definition 2.35. The forced system (2.8) is L_2-stable if for every $\eta_1 \geq 0$, every $\eta_2 \geq 0$, there is a number η_3 such that

$$\int_0^{\infty} \| \underline{x}(t;\ t_o,\ \underline{x}_o) \|^2 \, dt \leq \eta_3$$

for every initial condition $\underline{x}_o$ with $\| \underline{x}_o \| < \eta_1$ and every function $\underline{f}(t)$ with $\int_0^{\infty} \| \underline{f}(t) \|^2 \, dt < \eta_2$.

A possible definition of instability is the one obtained by negating that of stability. But it is more convenient to give a stronger form that includes the definition of 'complete instability' as found in Hahn [H 3(a), p 9]

Definition 2.36. The null solution of (2.1) or (2.4) is exponentially unstable if there exist positive constants $\varepsilon_1, \varepsilon_2$, such that for $t \geq t_o \geq 0$,

$$\| \underline{x}(t;\ t_o,\ \underline{x}_o) \| \geq \varepsilon_2 \, \|\underline{x}_o\| \, \exp(\varepsilon_1(t - t_o)) \tag{2.10}$$

Note that with $\varepsilon_1 = 0$, Definition 2.36 reduces to that of Hahn for complete instability. However, Yakubovich [Y 1(b)] uses complete instability in our sense of (2.10).

Definition 2.37. The forced system (2.8) is L_2-unstable if for every $\eta_3 > 0$, there exists a $\varepsilon > 0$ such that if $\|\underline{x}_0\| < \eta_1$ and for at least one $\underline{f}(t)$ with $\int_0^\infty \|\underline{f}\|^2 \, dt < \eta_2$,

$$\int_0^\infty e^{-\varepsilon t} \|\underline{x}(t;\, t_0,\, \underline{x}_0)\|^2 \, dt \geq \eta_3 \tag{2.11}$$

The definition of L_2-instability is different from the one found in literature [W 3(b)-(d)] but turns out to be more convenient in the proofs of instability theorems.

Remark 2.31. If (2.5) and (2.6) were to represent the same physical system, then it is obvious that a knowledge of the behaviour of $\sigma(t)$ enables us to deduce the behaviour of $\underline{x}(t)$ from (2.5a). For example, if $\sigma(t)$ is bounded, so is $\underline{x}(t)$, that is, all the components of $\underline{x}(t)$ are bounded. If $\sigma(t)$ tends to zero, so do the components of $\underline{x}(t)$. However, (2.6), as a separate entity, can represent systems which cannot be described by an ordinary differential equation. In this case, Definitions 2.31-2.37 have to be modified. Definitions 2.31, 2.32, 2.34, and 2.36 hold with $\underline{x}$ replaced by σ and the norm by modulus, if $|f_0(t)| \leq \gamma_1 \exp(-\gamma_0 t)$ for some positive constants γ_1 and γ_0. Furthermore, the integral equation corresponding to (2.8), but generalizing it, is

$$\sigma(t) = f_0(t) - \int_{t_0}^{t} g(t-\tau)k(\tau)\varphi(\sigma(\tau))d\tau + \int_{t_0}^{t} g(t-\tau)f(\tau)d\tau, \quad t \geq t_0 \geq 0,$$

where $f(t)$ is the input to the system. Definitions (2.33) and (2.35) carry over with the following replacements : σ for $\underline{x}$, f for $\underline{f}$ and modulus for the norm. Note that the effect of the initial condition $\sigma(t_0)$ has already been absorbed into $f_0(t)$.

In order to define stability and instability for the more general integral equation (2.7), we need the following :

$f(\cdot)$ is in $L_2(0,\infty)$, the linear space of real valued functions with the property

$$\int_0^{\infty} |f(t)|^2 \, dt < \infty$$

and equipped with the norm

$$\|f(\cdot)\| = \left(\int_0^{\infty} |f(t)|^2 \, dt\right)^{1/2} . \qquad (2.12)$$

Definition 2.38. The system represented by (2.7) is L_2-stable if, for all $f(\cdot)$ in $L_2(0, \infty)$, $v(\cdot)$ is also in $L_2(0, \infty)$ and

$$\|v\| \leq \text{Const.} \; \|f\| . \qquad (2.13)$$

This definition is consistent with the idea that finite 'energy' inputs must produce finite 'energy' outputs in stable systems.

Definition 2.39. The system represented by (2.7) is L_2-unstable if, for at least one function $f(\cdot)$ in $L_2(0,\infty)$, v is not in $L_2(0,\infty)$.

Remark 2.32. The definition of L_2-instability as given above is somewhat inconvenient to handle in establishing instability criteria for the system represented by (2.7). This is evident from the contributions of some authors [W 3(b)-(d)] [T 1] [S 5] who have to resort to proof by contradiction in establishing instability criteria. A better definition of instability is one which quantifies the instability property as in Definition 2.37 above. An extension of this definition to (2.7) is now given.

Definition 2.310. The system represented by (2.7)is L_2-**unstable** if, for every $\eta > 0$, there exists a $\varepsilon > 0$ such that for at least one input with $\|f\| < \eta_1$,

$$\|v \exp(-\varepsilon t)\|^2 \geq \eta \; \|v\| \cdot \|f\| . \qquad (2.14)$$

Remark 2.33. The definition for stability and instability mean absolute stability and absolute instability in the terminology of Lur'e [L 8] ,

Aizerman and Gantmacher [A 3] . This is because no reference has been made in these definitions to smallness or otherwise of $k(t)$ and $\varphi(\cdot)$, provided these functions belong to certain specified classes.

We now consider the basic linear time invariant feedback system (Fig.2.3) which plays a major role in our stability and instability studies.

Form (a) :

$$\left.\begin{aligned} &p(D)y + K\, q(D)y = 0 \\ \text{or, equivalently,}\quad &\frac{d\underline{x}}{dt} = A_0\, \underline{x} - K\, \underline{b}\, \underline{c}'\underline{x} \end{aligned}\right\} \tag{2.15}$$

Form (b) :

$$\sigma(t) = f_0(t) - K \int_0^t g(t - \tau)\sigma(\tau)d\tau \tag{2.16}$$

Form (c) :

$$\left.\begin{aligned} v(t) &= f(t) - K\, \sigma(t) \\ \sigma(t) &= \sum_{i=1}^{\infty} g_i v(t - \tau_i) + \int_0^{\infty} g(\tau)v(t - \tau)d\tau \end{aligned}\right\} \tag{2.17}$$

where K is assumed, for convenience, to take values in $[0,\infty)$, and $t \geq 0$. Any other range of variation for K can be converted to this by a suitable transformation (Exercise 2.2).

The next step in the stability and instability analyses of the nonlinear time varying feedback system is to endow the basic linear time invariant feedback system (Fig.2.3) with stability and instability. Then the fundamental question is : <u>How does the nonlinear time varying feedback gain affect the stability or instability of the system of Fig.2.3 ?</u>

2.32. Stability and instability assumptions on the system of Fig.2.3.

ASSUMPTION S1. The linear time invariant system described by (2.15) is asymptotically stable for every constant K in $[0,\infty)$. When $p(s)$ does not have zeros in the right half complex plane, the Nyquist plot of $G(j\omega)$ avoids the negative real axis. (See Desoer [D 4(a)]). When $p(s)$ has P zeros in the right half complex plane, the Nyquist plot of $G(j\omega)$ encircles the negative real axis P times+.

Note that Assumption S1 requires that all the eigenvalues of $(A_o - K\, \underline{b}\, \underline{c}')$ have negative real parts for K in $[0, \infty)$.

ASSUMPTION S2. Concerning (2.17), which includes (2.16) as a special case, we assume that $g(\cdot)$ is an element of $L_1(0,\infty)$, i.e.,

$$\int_0^\infty |g(t)|\, dt < \infty,$$

and there is a constant $\eta_o > 0$ such that $g(t)\exp(\eta_o t)$ is also in $L_1(0,\infty)$. Let $G_1(s)$ represent the Laplace transform of $\mathcal{G}$ defined by

$$G_1(s) = \sum_{i=1}^{\infty} g_i \exp(-s\,\tau_i) + \int_0^\infty g(t)\exp(-s\,t)dt. \tag{2.18}$$

Further, the system represented by (2.17) is L_2-stable for all K in $[0,\infty)$. Equivalently, the complex Nyquist plot of $G_1(s)$ satisfies the inequality

$$\inf_{\text{Re } s \geq 0} \left| 1 + K\, G_1(s) \right| > 0 \tag{2.19}$$

for all K in $[0,\infty)$.(See Desoer [D 4(a)] , Willems [W 3(b)], Davis [D 3(b)] and Saeks [S 1]).

ASSUMPTION U1. The linear time invariant system described by (2.15) is unstable for every constant K in $[0,\infty)$. When $p(s)$ has P zeros in the right half complex plane, the Nyquist plot of $G(j\omega)$ avoids the negative real axis or encircles the negative real axis P-1 or less number of times. When $p(s)$ does not have zeros in the right half complex plane, the Nyquist plot of $G(j\omega)$ encircles the negative real axis at least once.

Note that Assumption U1 requires that at least one eigenvalue of $(A_o - K\, \underline{b}\, \underline{c}')$ has a positive real part for all K in $[0,\infty)$. This may seen very restrictive with respect to K. However, if it is known that (2.15) is

---

+ For an equivalent statement in terms of a 'multiplier' function, see Chapter.3.

unstable for every constant K in a finite range, then by a suitable transformation the finite range can be converted to an infinite range (Exercise 2.2).

ASSUMPTION U2. The system described by (2.17), which includes (2.16) as a special case, is L_2-unstable for all constant K in $[0,\infty)$. In other words, inequality (2.19) is violated for all K in $[0,\infty)$. (See Willems [W 3(b)] and Davis [D 3(b)]).

2.4. Formulation of Stability and Instability Problems.

PROBLEM 2.41. The Generalized Stability Problem of Aizerman - Lur'e - Postnikov.

(i) For the system described by (2.1) or (2.4), with the nonlinear time varying gain satisfying the inequality

$$0 < \varepsilon \leq \frac{k(t)\varphi(\sigma)}{\sigma} < \infty \quad \text{for some } \varepsilon > 0 \text{ and all } t \geq 0 \text{ and } \sigma \neq 0, \qquad (2.20)$$

find, under Assumption S1, conditions, involving $G(j\omega)$ and $k(t)$, for exponential stability.

(ii) For the system described by (2.6), satisfying (2.20) and under Assumption S2 (specialized to (2.16)) find conditions involving $G(j\omega)$ and $k(t)$ for exponential stability.

(iii) For the system described by (2.7), satisfying (2.20) and under Assumption S2, find conditions involving $G_1(j\omega)$ and $k(t)$ for L_2-stability.

Remark 2.41. Problem 2.41 is a generalization of the stability problem as found in the work of Aizerman and Gantmacher [A 3] . It could also be termed as the problem of global exponential stability or exponential stability 'in the large'. The reasons for our choice of exponential stability in place of the normally used asymptotic stability [N 2] will be presented later (Sec. 2.41). It should be added here that two types of solution to Problem 2.41

(as also to other problems stated below) are possible : (a) Derivation of solution properties, such as boundedness, under Assumption S1 or Assumption S2 as the cases may be, without any additional restrictions. (b) Imposition of additional restrictions on $k(t)$, like bounds on it or its derivative, and on $\varphi(\cdot)$, like monotonicity or odd monotonicity, to assure exponential or L_2-stability, as the case may be. We deal mainly with the latter, and suggest the former as exercises. Note further that ε in (2.20) could as well be zero. But in such a case, $G(s)$ would have to satisfy the so-called 'stability-in-the-limit' conditions [A 3].

The next problem is the 'counterpart' of Problem 2.41.

PROBLEM 2.42. Instability Problem.

(i) For the system described by (2.1) or (2.4), satisfying (2.20) and under Assumption U1, find conditions involving $G(j\omega)$ and $k(t)$ for exponential instability.

(ii) For the system described by (2.6), satisfying (2.20) and under Assumption U2 (specialized to (2.16)), find conditions involving $G(j\omega)$ and $k(t)$ for exponential instability.

(iii) For the system described by (2.7), satisfying (2.20) and under Assumption U2, find conditions involving $G_1(j\omega)$ and $k(t)$ for L_2-instability.

Remark 2.42. The essence of the Nyquist criterion for linear time invariant feedback systems is that its general formulation does not make any assumptions concerning the internal dynamics but rather makes assumptions only on its input-output relation. In solving Problems 2.41 and 2.42, it is desirable to retain this elegance of the Nyquist criterion for nonlinear time varying feedback systems.

The following two problems, important as they are, do not seem to have been attempted in the literature.

PROBLEM 2.43. In the statement of Problem 2.41, replace Assumption S1 by Assumption U1, and Assumption S2 by Assumption U2.

PROBLEM 2.44. In the statement of Problem 2.42, replace Assumption U1 by Assumption S1, and Assumption U2 by Assumption S2.

Remark 2.43. Problems 2.43 and 2.44 certainly are meaningful and have some relevance. In Chapter 1, Sec.1.21(c), examples are given of linear time varying second order systems for which solutions to these problems are known from classical techniques. See also Kalman [K 1(a)].

Furthermore, an interesting problem has originated from the contributions of Pyatniskii [P 10(a)]. To formulate it, we need to write the equations for the linear time varying feedback system (Fig.2.4) which is a special case of the one given in Figs. 2.1 and 2.2

Form (a)
$$\left.\begin{aligned} & p(D)y + k(t)q(D)y = 0 \\ \text{or, equivalently,}\quad & \frac{d\underline{x}}{dt} = A_0\,\underline{x} - k(t)\underline{b}\,\underline{c}'\,\underline{x} \end{aligned}\right\} \tag{2.21}$$

Form (b) :
$$\sigma(t) = f_0(t) - \int_0^t g(t-\tau)k(\tau)\sigma(\tau)d\tau \tag{2.22}$$

Form (c) :
$$\left.\begin{aligned} v(t) &= f(t) - k(t)\sigma(t) \\ \sigma(t) &= \sum_{i=1}^{\infty} g_i v(t-\tau_i) + \int_0^{\infty} g(\tau)v(t-\tau)d\tau, \end{aligned}\right\} \tag{2.23}$$

all the equations holding for $t \geq 0$.

PROBLEM 2.45. Suppose that the system described by (2.21) is exponentially stable for all $k(t)$ taking values in $[\varepsilon, \infty)$ for some $\varepsilon > 0$, does it imply that the system (2.1) or (2.4) is also exponentially stable for all $k(t)\varphi(\cdot)$ satisfying inequality (2.20) ?

Corollary 2.451. Suppose that the system described by (2.22) is exponentially

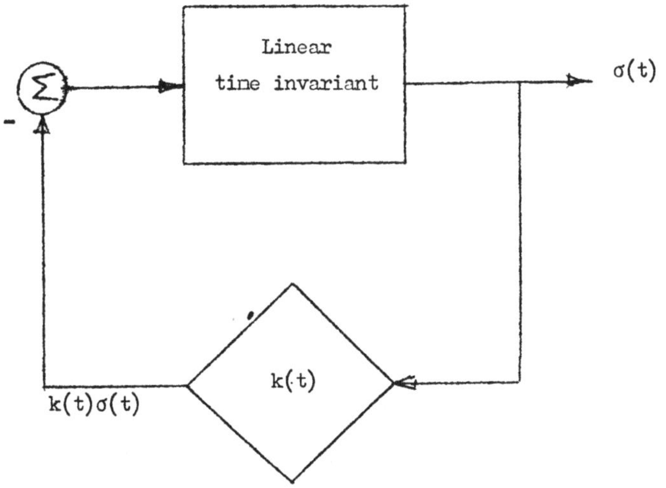

Fig.2.4(a). A linear time varying feedback system

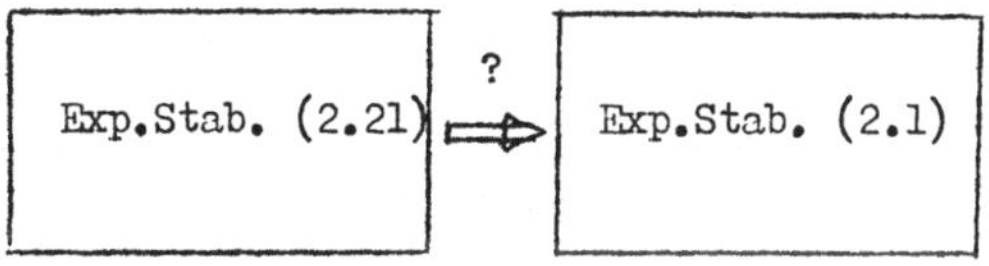

Fig.2.4(b). The Aizerman-Pyatnitskii Problem

stable for all $k(t)$ taking values in $[\varepsilon,\infty)$ for some $\varepsilon > 0$, does it imply that the system described by (2.6) is also exponentially stable for all $k(t)\varphi(\cdot)$ satisfying (2.20) ?

Corollary 2.452. Suppose that the system described by (2.23) is L_2-stable for all $k(t)$ taking values in $[\varepsilon,\infty)$ for some $\varepsilon > 0$, does it imply that the system described by (2.7) is also L_2-stable for all $k(t)\varphi(\cdot)$ satisfying (2.20) ?

Problem 2.45 is a generalization, to some extent, of the classic Aizerman problem which reads :

If Assumption S1 holds, does it imply that the nonlinear time invariant system

or, equivalently,

$$\left.\begin{aligned} p(D)y + \varphi(q(D)y) &= 0 \\ \frac{d\underline{x}}{dt} &= A_0\underline{x} - \underline{b}\,\varphi(\underline{c}'\underline{x}) \end{aligned}\right\} \text{ on } [0,\infty) \tag{2.24}$$

with $\varphi(\cdot)$ satisfying the inequality

$$0 < \varepsilon \leq \frac{\varphi(\sigma)}{\sigma} < \infty \tag{2.25}$$

for some $\varepsilon > 0$ and for all $\sigma \neq 0$, is also exponentially stable ? Note that Assumption S1 should be made a little stronger by requiring that (2.15) becomes unstable for $K < 0$. Only then the full force of the Aizerman problem becomes evident.

Remark 2.44. There is an important difference between the Aizerman problem and Problem 2.45. The difference lies in the fact that Assumption S1 can be easily verified using the familiar Nyquist or Routh - Hurwitz criterion, and involves necessory and sufficient conditions. If the Aizerman problem had an affirmative answer, then the absolute stability problem of time invariant nonlinear systems would reduce to a linear problem. In contrast with this, the exponential stability conditions for (2.21) are more difficult to establish, and, what is more important, even to this day have the character of sufficiency only. However, Problem 2.45 is interesting because the conclusions of Pyatniskii [P 10(a)] are different from those on the Aizerman problem: As is well known, the answer to the latter is negative [P 6], while the answer to Problem 2.45 as given by Pyatniskii is affirmative. See in this context [B 1]. If anything, the result of Pyatniskii exhibits a distinct property of time varying systems.

2.41. Reasons for the choice of Exponential Stability in Problem 2.41.

(i) It turns out that exponential stability conditions for (2.1) or (2.4) under Assumption S1 involve a certain global bound on the normalized rate of variation,(dk/dt/k). For L_2-stability stability of (2.7) under Assumption S2, a similar restriction on (dk/dt/k) appears. This prompts us to conjecture that exponential stability of (2.1) or (2.4) is equivalent to its L_2-stability (Definition 2.5). In fact, this has been established for the linear time varying system described by (2.21) [A 4]. Further, bounded-input bounded-output stability (Definition 2.6) of (2.21) is equivalent to its exponential stability [S 4] [L 3] [K 4,p 503].

(ii) Recently, stability analysis has been extended to include feedback systems of the type of Fig.2.2, the forward block ($\mathcal{G}$) containing time varying coefficients [B 12]. The main assumption made regarding $\mathcal{G}$ is that it is exponentially stable. Further, such an assumption is required in stability analysis of multiloop systems [M 8].

(iii) Exponential stability is a necessary and sufficient condition for the existence of quadratic forms[+] as Lyapunov function candidates[+] for (2.21) [B 9].

[+] See Sec.2.52. below.

(iv) The concept of exponential stability is important to a control engineer since it places a finite value on the settling time required for a perturbed motion to decay to a negligible value.

For other definitions of stability and inter-relations among them, see Sansone and Conti [S 3, p 485] , Yoshizawa [Y 3].

Remark 2.45. In the statement of Problem 2.42, exponential instability has been chosen as it is an exact counterpart of exponential stability. This type of instability is obviously stronger than actually needed in practice : The system could be practically unstable (Chapter 1, Sec.1.3(d)) even though it is not exponentially unstable. That is, the system solutions may become large enough after a lapse of time for small initial conditions even though they may not grow exponentially. Yakubovich [Y 1(b)] also uses exponential instability which he calls 'complete instability.' Note that Assumption U1 implies that (2.15) has solutions which grow exponentially and hence is exponentially unstable.

2.5. Methods used in Stability and Instability Analyses.

It is well known that (2.15) is asymptotically stable for all K in $[0,\infty)$ if and only if the roots of the algebraic equation

$$P_1(s) = s^n + p_{n-1} s^{n-1} + \cdots + p_o + K(q_n s^n + q_{n-1} s^{n-1} + \cdots + q_o) = 0$$

have negative real parts or the eigenvalues of $A_1 \triangleq (A_o - K \underline{b} \underline{c}')$ have negative real parts for all K in $[0,\infty)$.

Polynomial $P_1(s)$ having this property is called a Hurwitz polynomial, and matrix A_1, Hurwitz matrix. The equivalent interpretation in terms of the Nyquist plot of $G(j\omega)$ is known. Tne Hurwitz or the Nyquist criterion is inapplicable to the linear time varying system described by (2.21) except when the rate of variation of k(t) is very slow. See Rosenbrock [R 6] . If k(t) varies slowly compared to the natural frequency of oscillations of the linear time invariant system described by (2.15), the method of asymptotic expansions in powers of a small parameter, developed by N.M.Krylov and N.N.Bogoliubov [M 11] could be used for determining solutions of (2.21) and (2.1)

Leaving aside the case of time invariant linear system (2.15), the next type of linear system which has been extensively studied is the one described by (2.21) with a periodic $k(t)$. In this case, the classical result of Floquet [C 6] yields a foothold by asserting that every solution of (2.21) has the form

$$\underline{x}(t) = (\exp B\, t)\underline{r}(t)$$

where $\underline{r}(t)$ is a periodic vector with the same period as $k(t)$, and B is a constant matrix. This is an existence theorem which of itself furnishes no information as to the nature of B, even for a second order equation. For other results (of the existence type) on periodic coefficient differential equations, see Cesari [C 2] , Starzhinskii [S 9] . See also the recent article by R.S.Bucy [A 5(b), pp 65-94] on the evaluation of B via the Riccati equation.

For arbitrary $k(t)$, explicit solutions of (2.21) are unknown. The inference is that stability and instability analyses of (2.21), and hence of (2.1), cannot be based on assumed solutions forms. New techniques are needed to derive information concerning the nature of the solutions for (2.1) or (2.4), (2.6) and (2.7) directly from the properties of $k(t)$ and $\varphi(\cdot)$ without the intervention of explicit analytical solutions.

2.51. We could treat stability and instability problems in terms of the classical analysis of operators and their transformation properties. For instance, (2.6) could be written as the operator equation

$$(F\,\sigma)(t) = f_0(t), \tag{2.26}$$

where $$(F\,\sigma)(t) \triangleq \sigma(t) + \int_{t_0}^{t} g(t-\tau)k(\tau)\varphi(\sigma(\tau))d\tau \qquad (2.27)$$

The inverse of F, if it can be evaluated at all, leads naturally to the solution of (2.6). Statements about stability and instability of (2.6) are based on the properties of the inverse of F. Such an interpretation is at best as useful as the Floquet theory for linear periodic coefficient systems : Even for the linear time invariant system (2.15), it gives no explicit conditions for stability and instability for which the classical criteria of Nyquist have once again to be invoked. See Zames [Z 3(f)] , Willems [W 3(b)-(d)] , Damborg [D 1] , Damborg and Naylor [D 2] . The same comment holds for the inverse transformation approach as applied to (2.21) and (2.22) by Davis [D 3(a)] .

It should be added, however, that the inverse transformation approach may serve to provide a unifying point of view for the stability and instability analyses of general feedback systems. See Damborg [D 1] . In this context, an interesting subject for discussion has come up. Damborg [D 1] and Willems [W 3(b)-(d)] invoke causality+ and noncausality+ properties of the inverse operator for a statement of stability and instability conditions. As admitted by Willems [W 3(b)] , the steps needed to demonstrate causality and noncausality of operators are subtle and involved. In fact, even for the linear time invariant system (2.15), we have to resort to the Nyquist criteria for such a demonstration. Consequently, for more general equations of the

+ All physical systems are causal, i.e., the output only follows the input but does not precede it. See Zadeh and Desoer [Z 2] . A noncausal system is one in which the output precedes the input or, equivalently, the output predicts the input. See Chapter 4 and also, for instance, Willems [W 3(c)], Damborg [D 1] for mathematical definitions.

type (2.1), (2.6) or (2.7), it appears that no independent criteria can be given for causality and noncausality of the inverse operators.

In contrast with what has been related above, Skoog [S 5, p84], in the course of an instability analysis (Problem 2.42) of a special case of (2.7), declares that causality / noncausality and stability / instability are independent entities and should be treated as such. This reminds us of Bertrand Russell's epigram : Mathematics is the science in which we do not know what we are talking about, and do not care what we say about it is true.

We set aside general operator equations and concentrate on the specific feedback system of Figs.2.1 and 2.2. Further, we do not bring in considerations of causality and noncausality of inverses, while deriving explicit conditions for stability and instability. The explicit stability and instability conditions may very well act as a starting point for a separate demonstration of the existence and causality or noncausality of the inverse operator. Such a demonstration, which is believed to be of independent interest, may throw some light on a solution to the problem of stability and instability for more general systems using the concepts of causality and noncausality of inverse operators.

2.52. Classical and Modern Methods in Stability and Instability Analyses.

Classical results on stability for systems described by a differential equation (2.1) or (2.4) of order equal to or greater than 2, as found in Bellman [B 4(a)], Cesari [C 2], are derived from the Bellman - Gronwall lemma, Lemma 2.52 below, under the assumption that the "perturbation term", $k(t)\underline{b}\,\varphi(\underline{c}'\underline{x})$, is small in some sense. The perturbation term could be termed small if, for instance,

$$\frac{|\varphi(\underline{c}'\underline{x})|}{\|\underline{x}\|} \to 0 \quad \text{as} \quad \|\underline{x}\| \to 0$$

and (i) $|k(t)| \longrightarrow 0$ as $t \to \infty$; or

(ii) $\int_0^\infty |k(t)|\, dt < \infty$; or

(iii) $\int_0^\infty \left|\frac{dk}{dt}\right| dt < \infty$.

For the linear equation (2.21), a classic stability result due to Dini-Hukuhara [C 2] is as follows :

Lemma 2.51. If $k(t)$ is a measurable function and

$$\int_0^\infty |k(t) - K|\, dt < \infty$$

and the equation (2.15) has all solutions bounded on $[0,\infty)$, then (2.21) also has all solutions bounded.

Lemma 2.52. Let three real functions λ , w and u be defined in $[t_0, t_1]$ and continuous and let $\lambda(t) > 0$. Suppose the following inequality holds

$$u(t) \leq \int_{t_0}^{t} \lambda(\tau)u(\tau)d\tau + w(t), \quad t \in [t_0, t_1]. \tag{2.28}$$

Then $$u(t) \leq w(t) + \int_{t_0}^{t} \lambda(\tau)w(\tau) \exp\left(\int_{\tau}^{t} \lambda(\tau')d\tau'\right)d\tau$$

in $[t_0,t_1]$. Suppose w is differentiable, from inequality (2.28) it follows that

$$u(t) \leq w(t_0) \exp\left(\int_{t_0}^{t} \lambda(\tau)d\tau\right) + \int_{t_0}^{t} \exp\left(\int_{\tau}^{t} \lambda(\tau')d\tau'\right) \frac{dw}{d\tau}\, d\tau .$$

Further, if w is a constant, then from inequality (2.28) there follows

$$u(t) \leq w \exp\left(\int_{t_0}^{t} \lambda(\tau)d\tau\right)$$

Proof : See [H 4, p 7].

We are concerned not with a small perturbation but with large perturbations or, to use the terminology of Barbashin and Krasovskii [B 3], stability and instability of systems 'in the large.'

Note that classical results fail to indicate stability of (2.4) in the absence of asymptotic stability of

$$\frac{d\underline{x}}{dt} = A_0 \underline{x} \tag{2.29}$$

(where A_0 is as defined in Sec.2.2) and instability of (2.4) in the absence of complete instability of (2.29). Note further that the integrability of $|k(t) - K|$ in Lemma 2.51 excludes the case of $k(t)$ slowly varying with respect to K or some other constant. The point of view adopted by us is that whatever stability and instability criteria are derived for (2.21), (2.1), (2.6) or (2.7) should also apply to the case $k(t) = \text{constant}$, a requirement not met by some of even the latest instability criteria [S 5] [T 1] [B 18] [W 3(d)].

Furthermore, it is found that the assumption of a nonnegative $\lambda(t)$ in Lemma 2.52 is very restrictive and is responsible for the conservative stability estimates on the perturbation term of (2.1) or (2.4) as derived in [B 4(a)] and others. But for instability (Problem 2.42), note that there is no counterpart of Lemma 2.52. The classical instability results, although based on the premise that the perturbation term is small in some sense, are established by different means reminiscent of Lyapunov's method.+ See, for instance, Bellman [B 4(a), pp 88-89].

More important methods in the present day qualitative theory of differential equations of the type (2.1) or (2.4) involve integration of differen-

---

+ See below

tial inequalities. See Lakshmikantham and Leela [L 1] for a comprehensive treatment of differential inequalities. For integral equations of the type (2.7), Cauchy - Schwarz inequality in L_2-space (along with integration of differential inequalities) plays an important role.

Modern stability theory is founded on the contributions of

(a) Lyapunov [L 10] as improved by Corduneanu [C 8] (see also Krasovskii [K 7, pp 56-57], Melnikov [M 7]) for systems described by differential equations ;

(b) Popov [P 8(a),(b)] for systems of the type (2.6) ; and

(c) Zames [Z 3(a),(b)] and Sandberg [S 2(a)] for systems of the type (2.7).

We now give the main ideas of the three contributions as applied to the non-linear time varying feedback system, Figs.2.1 and 2.2 .

2.52(a) Lyapunov - Corduneanu.

In this method, which treats stability and instability problems directly but in a nonconstructive way, the so-called differentiation of a suitably chosen function, termed here as a Lyapunov - Corduneanu function candidate, along the solutions of (2.1) or (2.4) is used.

Even though application of this method, unlike the classical methods, is not limited by the artificial requirement that (2.15) or (2.29) be completely stable / unstable, we operate under Assumption S1 and Assumption U1 for stability and instability analyses (Problems 2.41 and 2.42) of (2.4) respectively. No solution seems to be available for Problems 2.43 and 2.44 using the Lyapunov - Corduneanu method. For equations different in form from (2.4), the method is definitely limited by practical matters related to the construction or discovery of suitable Lyapunov - Corduneanu candidate functions.

Let $V(\underline{x},t) = V(x_1,x_2,\cdots, x_n,t)$ be some differentiable function of variables $x_1,x_2, \cdots , x_n$ and t defined in a certain domain. Its time derivative with respect to (2.4) is defined by means of the formula

$$\left.\frac{dV(\underline{x},t)}{dt}\right|_{(2.4)} \triangleq \frac{\partial V(\underline{x},t)}{\partial x_1} x_2 + \frac{\partial V(\underline{x},t)}{\partial x_2} x_3 + \cdots + \frac{\partial V(\underline{x},t)}{\partial x_{n-1}} x_n$$
$$+ \frac{\partial V(\underline{x},t)}{\partial x_n} \left\{ - p_o x_1 - p_1 x_2 - \cdots - p_{n-1} x_n - k(t)\varphi(q_o x_1 + q_1 x_2 + \cdots + q_m x_m) \right\}$$
$$+ \frac{\partial V(\underline{x},t)}{\partial t} . \tag{2.30}$$

In the proof of stability and instability theorems, a major role is played by positive definite quadratic forms. A quadratic form is a function $W(\underline{x})$ determined by the formula

$$W(\underline{x}) = \sum_{i,j=1}^{n} w_{ij}\, x_i\, x_j \quad , \quad w_{ij} = w_{ji} \text{ being real numbers.}$$

The quadratic form is positive definite if, for all $x \neq 0$, we have $W(\underline{x}) > 0$. For any positive definite quadratic form $W(\underline{x})$, it is always possible to choose two numbers ν_1 and ν_2 such that the inequality

$$\nu_1 \|\underline{x}\|^2 \leq W(\underline{x}) \leq \nu_2 \|\underline{x}\|^2 \tag{2.31}$$

holds for an arbitrary vector $\underline{x}$ [B 4(c)] [G 1] .

$W(\underline{x})$ is said to be negative definite if $(-W(\underline{x}))$ is positive definite. $W(\underline{x})$ is said to be positive semi-definite is $W(\underline{x}) \geq 0$ for all $\underline{x} \neq \underline{0}$. $W(\underline{x})$ is said to be negative semi-definite if $(-W(\underline{x}))$ is positive semi-definite.

For the linear time varying system (2.21), we use, in Chapter 3, a quadratic form of the type

$$V(\underline{x},t) \triangleq V_1(\underline{x}) + k(t)V_2(\underline{x}) \tag{2.32}$$

where $V_1(\underline{x})$ is a positive definite quadratic form, $V_2(\underline{x})$ is a positive semi-definite quadratic form. Note that inequality (2.31) is satisfied by $V(\underline{x},t)$.

For the nonlinear time varying system (2.1) or (2.4), we use, in Chapter 3, a function of the type

$$W(\underline{x},t) = V_1(\underline{x}) + k(t)e^{-2\alpha\tau} \int_0^t e^{2\alpha\tau} \varphi(\underline{c}'\underline{x})\underline{\gamma}'\underline{x} \, d\tau \tag{2.33}$$

where $V_1(\underline{x})$ is a positive definite quadratic form, and the integral term is constructed as to be nonnegative by a proper choice of α and the constant vector $\underline{\gamma}$. $W(\underline{x},t)$ defined in (2.33) satisfies the inequality

$$\nu_1 \|\underline{x}\|^2 \leq W(\underline{x},t) \leq \nu_2 \|\underline{x}\|^2 + \nu_3 \, e^{-2\alpha t} \int_0^t e^{2\alpha\tau} \|\underline{x}\|^2 \, d\tau$$

for some positive constants ν_1 , ν_2 and some nonnegative constant ν_3.

Construction of $V_1(\underline{x})$ and $V_2(\underline{x})$ appearing in (2.32) and (2.33) will be accomplished in Chapter 3 using a technique due to Brockett [B 16(b)].

The time derivative of $V(\underline{x},t)$ with respect to (2.21) and that of $W(\underline{x},t)$ with respect to (2.1) or (2.4) are then written in the form of a differential inequality :

$$\frac{dv}{dt} \leq \lambda(t) v , \quad t \geq t_o \geq 0 \tag{2.34}$$

where $\lambda(t)$, expressible in terms of $k(t)$ and (dk/dt), is a real valued function on $[t_o,\infty)$ with $t_o \geq 0$. Integrating (2.34) and restricting the behaviour of $\lambda(t)$ suitably, exponential stability **conditions** for (2.1)and (2.21) are derived.

For instability analysis (Chapter 6) of (2.1) and (2.21) (Problem 2.42), Lyapunov - Corduneanu function candidates are suitably constructed, and their derivatives with **respect** to an auxiliary system corresponding to (2.1) and (2.21) are written in the form of the differential inequality

$$\frac{dv}{dt} \geq \mu(t)v , \quad t \geq t_o \geq 0 \tag{2.35}$$

where $\mu(t)$, expressible in terms of $k(t)$ and (dk/dt), is a real valued function on $[t_0,\infty)$ with $t_0 \geq 0$. Integrating (2.35) and restricting the behaviour of $\mu(t)$ suitably, exponential instability conditions for (2.1) and (2.21) are derived.

2.52(b) Popov.

In this method,(Chapter 4) applicable to (2.6), a quadratic functional of the form

$$\rho(T) = \int_0^T k(t)\varphi(\sigma(t))(z(t) * \sigma(t))dt \tag{2.36}$$

where $*$ denotes convolution and $z(t)$ is real valued function on $(-\infty,\infty)$ but belonging to the class of L_1 functions so chosen as to make $\rho(T)$ positive for all $\sigma(t) \neq 0$, is evaluated with respect to (2.6) :

$$\rho(T)\Big|_{(2.6)} = \int_0^T k(t)\varphi(\sigma(t))(g(t) * z(t) * \varphi(\sigma(t))\ dt \tag{2.37}$$

Parseval's theorem in the theory of Fourier transforms is invoked to derive,first,positivity conditions for (2.37) and, then,in combination with (2.36), an integral inequality of the form

$$\nu_1\sigma^2(T) + \nu_2\int_0^T \sigma^2(t)dt \leq \nu_3\sigma^2(0) \tag{2.38}$$

for some positive constants ν_1, ν_2 and ν_3 and for all $T > 0$. From this inequality, asymptotic stability of (2.6) can be established. Exponential stability of (2.6) requires some minor additional restrictions. Stability analysis of the linear system (2.21) is included in the above treatment. Comparison of (2.38) with the Bellman - Gronwall inequality (2.28) shows why we get conservative results by using the latter. Observe that ν_2 should be negative for applying Lemma 2.52.

Derivation of instability conditions (Problem 2.42) using Popov's method does not seem to be known in literature (see the statement of open problems in the last section in Brockett and Lee [B 18]. This could be treated as a special case of the instability analysis, using the converse Schwarz inequality [M 12], of systems governed by (2.7).

2.52.(c) Zames - Sandberg.

The method of Zames - Sandberg (Chapter 5), applied to systems described by (2.7), uses essentially the same idea as in Popov's method, the difference being that $\rho(T)$ defined in (2.36) is to be treated as an inner product in L_2-space. After evaluating $\rho(T)$ with respect to (2.7), Parseval's theorem and the Cauchy - Schwarz inequality are invoked to derive an inequality of the form

$$\gamma_1 \|\sigma\|^2 \leq \gamma_2 \|\sigma\| \cdot \|f\| + \gamma_3 \qquad (2.39)$$

for some positive constants γ_1, γ_2 and γ_3, from which L_2- stability of (2.7) is concluded.

As regards L_2-instability (Chapter 7)(Problem 2.42), a suitably chosen quadratic functional is evaluated with respect to an auxiliary system corresponding to (2.7). Application of Parseval's theorem and the converse Schwarz inequality then leads to an inequality of the form

$$\| \sigma \exp(-\varepsilon t) \|^2 \geq \eta \; \|\sigma\| \; \|f\| \qquad (2.40)$$

for some positive constants ε and η and some $f(\cdot)$ belonging to a non-trivial subspace of $L_2(0,\infty)$. Inequality (2.40) implies in particular that there are outputs which do not belong to $L_2(0,\infty)$. What is, however, more significant is that (2.40) gives a lower bound on the norm of solutions of (2.7).

2.53. General Comments.

(i) The problem of existence and uniqueness of solutions for the feedback system of Figs. 2.1 and 2.2 is outside the ambit of the monograph. See Coddington and Levinson [C 6], Halanay [H 4] and Willems [W 3(c)] for these considerations. As explained in Hahn [H 3(a),p 9], a slight modification of the stability and instability definitions becomes necessary if the assumptions about the right hand side of (2.4) are weakened so that only existence but not the uniqueness of the solutions is guaranteed. The distance of the motion' from the null solution has to be defined, while considering stability, by a supremum type of norm and, while dealing with instability, by an infimum type of norm. For equations more general than (2.7), Zames [Z 3(a)(e)(f)] indicates that stability and instability are to be treated as different from considerations of existence and uniqueness of solutions. As is well known, existence and uniqueness of solutions can be established by iterative procedures whereas stability and instability cannot.

The point of view adopted here is that whatever solutions exist have the properties of stability and instability established by the noniterative methods. It is quite likely that stability conditions along with some minor restrictions will guarantee existence and uniqueness of solutions of the system under consideration [Z 5]. Similarly instability conditions with or without other restrictions may give an indication of existence and nonuniqueness or nonexistence of solutions. However, in the inetrest of simplicity, it is convenient to assume for all the system equations under our consideration existence and uniqueness of solutions.

(ii) In contrast with the geometric arguments for the application of the Lyapunov method as found in the excellant monograph of LaSalle and Lefschetz [L 4], we draw upon integration of differential inequalities.

(iii) It turns out that stability and instability results obtained by the three methods for the feedback system of Figs.2.1 and 2.2, are almost identical. The only noticeable difference is that, for stability, a certain type of restriction on $k(t)$ seems to be derivable only by the Popov and

Zames - Sandberg methods, and not by the Lyapunov - Corduneanu method. This difference and the inadequacy of all the results as applied to the stability and the instability analysis of the Mathieu or the Hill equation are significant motivation for further work.

(iv) A characteristic distinction between the Soviet followers of Lyapunov and research workers outside the Soviet Union is that stability and instability properties of (2.4) are established by the latter in an enlarged state space in which the original state space is invariantly embedded.

2.6. Conclusions.

Nonlinear time varying coefficient differential and integral equations could be treated as made up of simpler sets of equations of which the feedback system model of Figs.2.1 and 2.2 forms the prototype. Stability and instability problems centred on this model are formulated along with standard and nonstandard assumptions. Two problems out of these are tractable : The generalised Aizerman - Lur'e - Postnikov problem and its exact counterpart. The other problems are open.

The main ideas of the three modern methods available for stability analysis are outlined. The instability versions of these, which are not so well known, are given in brief. Considerable preliminary work remains before these ideas can be put to actual use. The remaining chapters deal with these details.

EXERCISES

2.1.(a) Show that the system of Fig.2.5 with $G'(s) = q'(s)/p(s)$ where

$$q'(s) = s^n + q_{n-1}s^{n-1} + \cdots + q_o ,$$

$p(s)$ is defined by (2.2) and $\varphi(\cdot)$ as in Sec.2.2(a), can be reduced to the system of Fig.2.6 in which $\varphi'(\cdot)$ now satisfies a finite sector inequality. Find the relation between y and y' , φ and φ' . Repeat the exercise for

$$q'(s) = \gamma(s^n + q_{n-1}s^{n-1} + \cdots + q_o)$$

where γ is a constant. Under what conditions is monotonicity of $\varphi(\cdot)$ in both the cases retained after transformation ?

(b) Replace $\varphi(\cdot)$ by $k(t)$ in Fig.2.5 and $\varphi(\cdot)$ by $k'(t)$ in Fig.2.6. Repeat part (a).

(c) Replace $\varphi(\cdot)$ in Fig.2.5 by $\varphi(\cdot)k(t)$ and $\varphi'(\cdot)$ in Fig.2.6 by $\varphi'(\cdot)k'(t)$. Repeat part (a).

2.2 (a) The linear time varying system of Fig.2.7 with $G(s)$ defined in Sec.2.2 and $k_1(t)$ satisfying the inequality

$$\eta_1 < k(t) < \eta_2 \quad \text{for all } t \text{ in } [0,\infty)$$

where η_1 and η_2 are two known constants and $\eta_1 < \eta_2$, is to be transformed to the system of Fig.2.8 with $k(t) = ((k_1(t) - \eta_1)/(\eta_2 - k_1(t))$. Show that $G_1(s) = (1 + \eta_2 G(s))/(1 + \eta_1 G(s))$. Find the relation between y' and y. Note that $k(t)$ takes values in $(0,\infty)$ for all t in $[0,\infty)$.

(b) Replace $G(s)$ of Fig.2.7 by $G'(s)$ of Fig.2.5 and repeat part (a).

(c) Replace $k_1(t)$ of Fig.2.7 by $\varphi_1(\cdot)k_1(t)$ satisfying the inequality

$$\eta_1 < (\varphi_1(\sigma)k_1(t)/\sigma) < \eta_2 \quad \text{for all } \sigma \neq 0 \text{ and } t \text{ in } [0,\infty)$$

for given constants η_1 and η_2 with $\eta_1 < \eta_2$. Repeat part (a).

2.3.. What are the differences, if any, between the system of Fig.2.2 and the various systems of Fig.2.9 as regards stability and instability properties of the loop variables ? Derive the relations, if they exist, between the output variable of Fig.2.2 and those of Fig.2.9. Note that for the last two systems

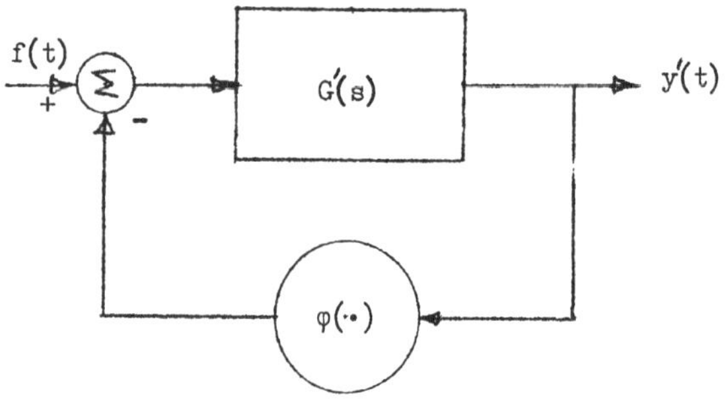

Fig.2.5

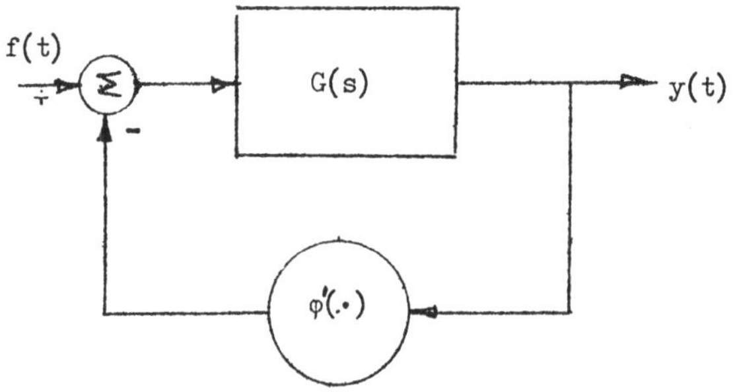

Fig.2.6

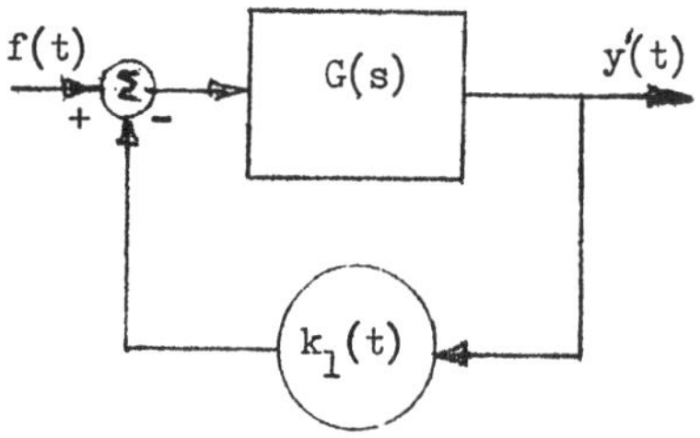

Fig.2.7

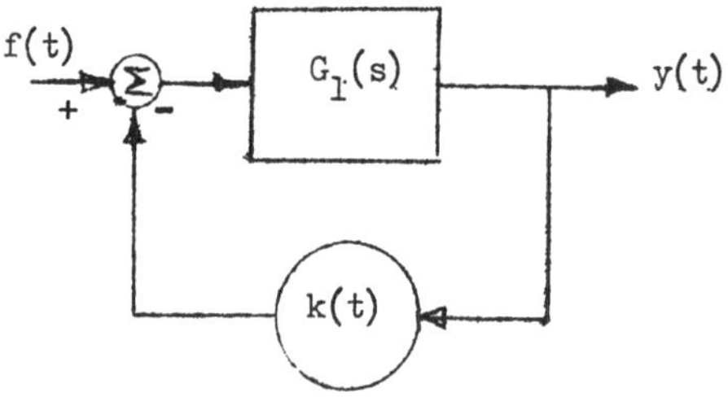

Fig.2.8

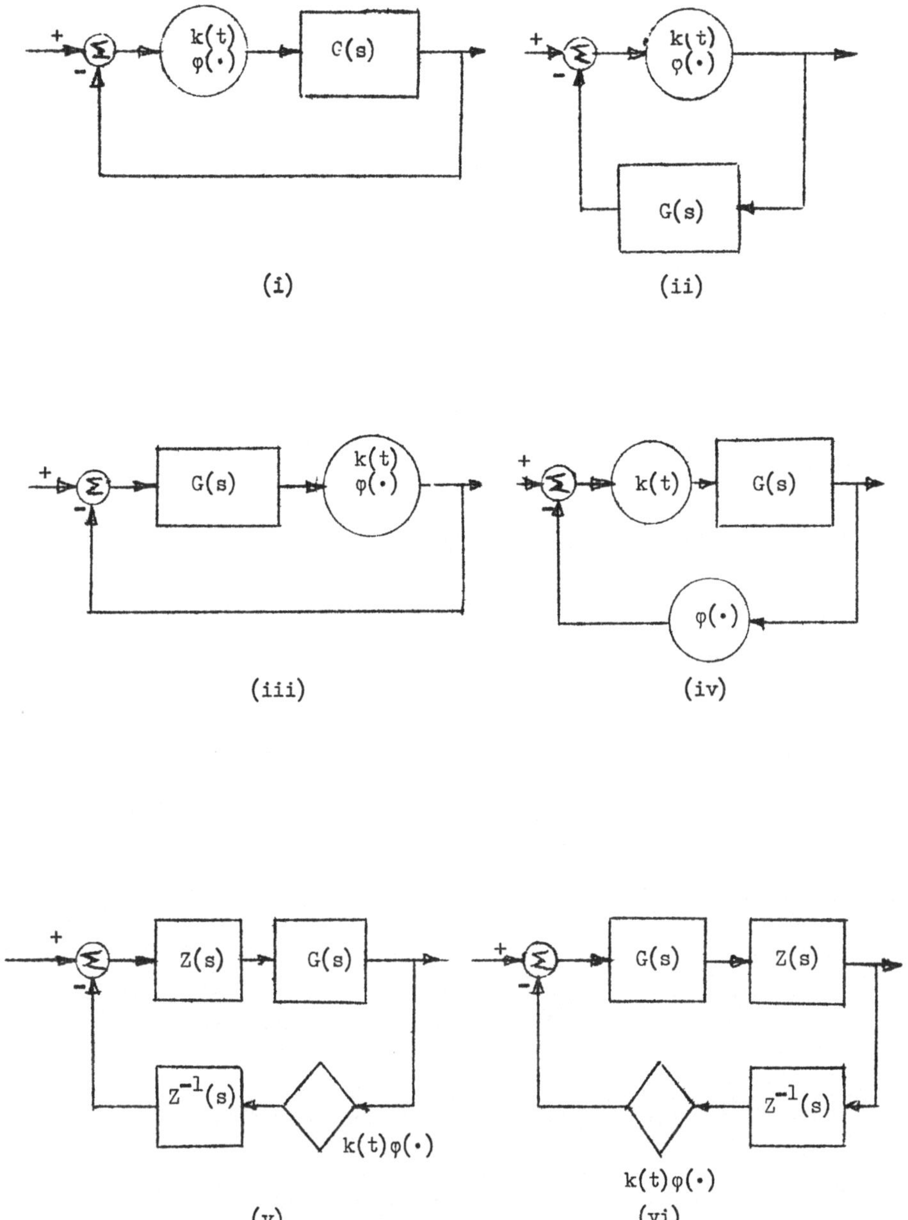

+
-
Σ
k(t)
φ(·)
G(s)
(i)
k(t)
φ(·)
G(s)
(ii)
G(s)
k(t)
φ(·)
(iii)
k(t)
G(s)
φ(·)
(iv)
Z(s)
G(s)
Z^-1(s)
k(t)φ(·)
(v)
G(s)
Z(s)
Z^-1(s)
k(t)φ(·)
(vi)

Fig.2.9

of Fig.2.9, Z(s) and $Z^{-1}(s)$ should be such that no instability is introduced into the system. That is, for a rational Z(s), the poles and zeros of Z(s) should be to the left of the imaginary axis in the complex plane. Now suppose that the poles and/or zeros of a rational Z(s) do exist on the right of the imaginary axis in the complex plane. Try any subterfuge for converting the stability problem of the last two systems of Fig.2.9 with rational Z(s) into that of system freed from these poles and/or zeros of Z(s). Suppose Z(s) were not rational, what should one do ? See Zames and Falb [Z 4] .

2.4. A linear system more general than the one considered in Sec.2.2, is for instance described by

$$\frac{d\,\underline{x}}{d\,t} = A(t)\,\underline{x}(t) - k(t)\,\underline{b}\,\underline{c}'\underline{x}(t) \tag{2.41}$$

where A(t) is time varying n x n matrix. Under suitable conditions of controllabiliy, Ramar and Ramaswamy [R 2] develop a transformation to reduce the system (2.41) to a scalar differential equation. Solve the inverse problem of delineating the classes of systems of the type (2.41) equivalent to (2.1). That is, find the classes of all matrices A(t) such that the system (2.41) is equivalent to (2.1). Note that for stability equivalence, the transformation matrix should possess special properties originally due to Lyapunov [Z 2] .

2.5. Consider the system, first discussed by Marcus and Yamabe [M 5] ,

$$\frac{d\,x_1}{d\,t} = (-1 + a\cos^2 t)x_1 + (+1 - a\sin t\cos t)x_2$$

$$\frac{d\,x_2}{d\,t} = (-1 - a\sin t\cos t)x_1 + (-1 + a\sin^2 t)x_2$$

with $a > 0$. Verify that the state transition matrix is given by

$$\Phi(t,0) = \begin{bmatrix} (\exp(a-1)t)\cos t & (\exp(-t))\sin t \\ -(\exp(a-1)t)\sin t & (\exp(-t))\cos t \end{bmatrix}$$

Find the eigenvalues of the system matrix. Compare the conclusions on stability made first on the basis of the transition matrix and next from the eigenvalues . What do you infer from the anomaly ?

2.6. Consider the system discussed by Wu [W 7(b)] ,

$$\frac{d x_1}{d t} = (-5\cdot 5 + 7\cdot 5 \sin 12t)x_1 + 7\cdot 5(\cos 12t)x_2$$

$$\frac{d x_2}{d t} = 7\cdot 5(\cos 12t)x_1 - (5\cdot 5 - 7\cdot 5 \sin 12t)x_2 .$$

Verify that the state transition matrix is given by

$$\Phi(t,0) = \begin{bmatrix} \phi_{11}(t) & \phi_{12}(t) \\ \phi_{21}(t) & \phi_{22}(t) \end{bmatrix}$$

where $\phi_{11}(t) = 0\cdot 5\exp(-t)(\cos 6t + 3\sin 6t) + 0\cdot 5\exp(-10t)(\cos 6t - 3\sin 6t)$

$\phi_{21}(t) = 0\cdot 5 \exp(-t)(3\cos 6t - \sin 6t) - 0\cdot 5\exp(-10t)(3\cos 6t + \sin 6t)$

$\phi_{12}(t) = (\exp(-t)(\cos 6t + 3\sin 6t)/6) - \exp(-10t)(\cos 6t - 3\sin 6t)/6$

$\phi_{22}(t) = (\exp(-t)(3\cos 6t - \sin 6t)/6) + \exp(-10t)(3\cos 6t + \sin 6t)/6.$

Repeat Ex.2.5.

2.7. Consider the equation $(dy/dt) = (-y + y^2)$ for $t \geq 0$. Show that solutions near the origin tend to the origin as $t \to \infty$, while solutions far away from the origin tend to infinity in finite time. What do you conclude about the applicability of results meant for (2.1) ?

2.8. Analyse the behaviour of solutions of the equation $(dy/dt) = (y - y^2)$ for $t \geq 0$. Compare with the conclusions of Ex.2.7.

2.9. Find the linearised equation for

$$(d^2y/dt^2) + \varphi(y)(dy/dt) + (\omega^2 + \varepsilon \cos t)y = 0 \quad \text{for} \quad t \geq 0,$$

assuming that $\varphi(0) = K > 0$. Find condition/conditions on the nonlinear function $\varphi(\cdot)$ for the classical stability/instability results to hold under suitable assumptions on the behaviour of the linearised equation.

CHAPTER 3

EXPONENTIAL STABILITY OF FEEDBACK SYSTEMS IN DIFFERENTIAL FORM

3.1. Introduction. In this chapter we apply the Lyapunov-Corduneanu theorem (Lemma 3.34 and Corollary 3.341) to derive exponential stability criteria for systems described by

$$p(D)y + k(t)\varphi(q(D)y) = 0 \quad \text{on the interval } [0,\infty) \qquad (2.1)$$

under assumptions set forth in Chapter 2. We consider the linear system described by

$$p(D)y + k(t)q(D)y = 0 \quad \text{on the interval } [0,\infty) \qquad (2.21)$$

separately. In brief, we attempt to resolve the question of exponential stability for (2.1) and (2.21) under suitable assumptions. That is, we deal with Problem 2.41 of Chapter 2.

As explained in Sec.2.52, we choose a suitable quadratic form (Sec. 3.3 and Sec.3.4) for the linear system (2.21), and a suitable quadratic form plus integral of the nonlinearity (Sec.3.5) for the nonlinear system (2.1), and evaluate their time-derivatives along the solutions of (2.21) and (2.1) respectively. The time-derivatives so evaluated are written in the form of a differential inequality (2.34) from which conditions for the exponential stability of (2.21) and (2.1) are obtained.

We seek open loop conditions for the closed loop stability of the system under consideration (Fig.2.1). It is therefore desirable that the exponential stability conditions on $\mathcal{G}$ be expressed in terms of the frequency characteristic $G(j\omega)$, and those on $k(t)\varphi(\cdot)$ be easily verifiable. Concerning the former, we employ Brockett's method [B 16(b)] of path independent integrals (Sec.3.3) for generating suitable quadratic forms. As regards the latter, the condition on $k(t)$ takes the form of a certain global bound on the normalised rate of variation $(dk/dt/k)$; other than the choice of a suitable class of nonlinearities, no more restrictions need be imposed on $\varphi(\cdot)$.

For the linear system (2.21), the stability requirement on $\mathcal{G}$ reduces to the well known Nyquist stability condition on $G(j\omega - \beta)$ for some constant $\beta > 0$. For the nonlinear system (2.1), such a statement of the stability condition on $\mathcal{G}$ is possible if $\varphi(\cdot) \in C$ or C_m. As is to be expected, exponential stability conditions for the system (2.1) involve a trade-off among (i) the class of nonlinearities ; (ii) normalized rate of variation of $k(t)$; and (iii) the phase angle characteristic of $G(j\omega)$.

Comparison with the results found in literature is made (Sec.3.4(b) and Sec. 3.6). Remarks interspersed in the body of the chapter, and exercises at the end are meant to show that many basic problems of stability remain unsolved.

3.2. Preliminaries. The following concepts from network theory are needed : Causal function, anticausal function, noncausal and positive real function. Why they are needed will be explained below (Lemma 3.22). The function $Z(s)$, which turns up frequently in stability analysis, denotes a complex valued function of a complex variable s. In the present chapter it is restricted to be a rational function of s.

Definition 3.21. $Z(s)$ is called a causal (anticausal) function if the inverse Laplace transform of $Z(s)$, i.e., the impulse response of the system characterized by $Z(s)$, is nonzero for $t \geq 0$ $(t \leq 0)$ and zero for $t < 0$ $(t > 0)$. The system is also said to be nonanticipative (anticipative).

$Z(s)$ is said to be noncausal if it contains both causal and anticausal terms (Exercise 3.1).

Definition 3.22. $Z(s)$ is called a positive real (strictly positive real) function of the argument s, if $Z(s)$ is real for real values of s and for $\text{Re } s > 0$ (where Re denotes the real part) is analytic and satisfies the inequality $\text{Re } Z(s) \geq 0$ (>0).

A positive real function is necessarily causal.

The following lemma concerning a rational $Z(s)$ will be used in the generation of positive definite quadratic forms.

Lemma 3.21. A real function of a complex variable $Z(s) = m(s)/n(s)$, where $m(s)$ and $n(s)$ are finite polynomials in s, is positive real if and only if
(i) $n(s) + m(s)$ has no zeros in the closed right half plane $\mathrm{Re}\, s \geq 0$, and
(ii) $\mathrm{Re}\, Z(j\omega) \geq 0$ for all real ω.

Proof. See Weinberg and Slepian [W 1].

Corollary 3.211. If hypothesis (ii) in Lemma 3.21 is replaced by
(ii)' $\mathrm{Re}\, Z(j\omega - \varepsilon) \geq 0$ for all real ω and for some real constant $\varepsilon > 0$, then $Z(s)$ is strictly positive real.

Remark 3.21. A network with positive and time invariant RLC elements possesses a positive real driving point impedance function. An RC-network has an impedance function of the form $\sum_i \alpha_i/(s + \beta_i)$ where α_i and β_i are constants with $\alpha_i > 0$ and $\beta_i \geq 0$ for all $i = 1,2,3,\cdots$. An RL-network has an impedance function of the form $\gamma_o + \gamma_1 s + \sum_i \alpha_i/(s + \beta_i)$ where γ_o, γ_1, α_i and β_i are constants with γ_o and γ_1 nonnegative, α_i nonpositive and β_i positive for all $i = 1,2,3,\cdots$. An LC-network has an impedance function of the form $\gamma_1 s + (\gamma_o/s) + \sum_i (\alpha_i s/(s^2+\beta_i^2))$ where γ_1, γ_o, α_i and β_i are nonnegative constants for all $i = 1,2,3,\cdots$. The arguments of these network functions (with s replaced by $j\omega$), which are bounded by $\pi/2$ and $-\pi/2$, play an important role in the statement of stability conditions.

Remark 3.22. Suppose $Z(s)$ is not positive real, then it is quite likely that $Z(s) + (1/K_1)$ is positive real for some $K_1 > 0$. If this is still not the case, it is possible that $(Z(s) + (1/K_1)) / (Z(s) + (1/K_2))$ is positive real for some positive constants K_1 and K_2 with $K_1 < K_2$. In these cases, it is desirable to express positive realness conditions in terms of the frequency plot of $Z(j\omega)$. See Exercise 3.4.

An alternative characterization of positive real functions is in the time domain : It is well known from network theory that if $Z(s)$ is a positive real function, and $z(t)$ is the inverse transform of $Z(s)$, then for any function $f(t)$

in the class of continuous real valued functions, the functional

$$\rho(T) = \int_0^T f(t) \left(\int_0^{\tau} z(t-\tau) f(\tau) d\tau \right) dt \geq 0 \tag{3.1}$$

for all $T > 0$. The system is said to be passive.

A relevant question at this point would be : How does a positive real function come into picture while analysing the stability of a nonlinear time varying system ? And why not a nonpositive real function ?

Part of the answer lies in the fact that, until 1961 when Popov's [P 8(a)] outstanding paper appeared, the Lyapunov-based results (for the stability of nonlinear systems) of the predominantly Russian school headed by Lur'e and Aizerman did not resolve, in simple terms, the problem of finding suitable candidates as Lyapunov functions and hence the stability conditions were not as elegant as the Nyquist criterion. Popov [P 8(a)] derived, in a way different from the Lyapunov method, absolute stability conditions for the time invariant nonlinear system (2.24) with $\varphi(\cdot) \in C$. These conditions were expressed in terms of a 'multiplier' function $(1 + \alpha s)$, where constant $\alpha \geq 0$, such that $(1 + \alpha s)G(s)$ is positive real or $\mathrm{Re}\,(1 + \alpha j\omega)G(j\omega) \geq 0$ for all real ω. Significantly, an explicit construction of a Lyapunov function candidate is avoided. Further, the absolute stability condition for the time invariant nonlinear system (2.24) is given in terms of a frequency domain inequality reminiscent of the Nyquist criterion. But the question which comes up is : Can we express the Nyquist stability criterion in terms of a multiplier function?Naturally, a further question would be : Under what conditions can these be generalized for guaranteeing the stability of time varying nonlinear systems ?

The next lemma concerns the interpretation of the Nyquist criterion in terms of a multiplier function. This lemma serves to show the gap between the Popov criterion and the Aizerman conjecture for time invariant nonlinear systems (Sec.2.4).

Assumption 3.21. There are no poles of $G(s)$ in the closed half plane

Re s $\geq$ $-\beta_{sh}$, and the frequency plot of $G(j\omega - \beta_{sh})$ with ω in $(-\infty,\infty)$ does not intersect the negative real axis including the origin.

Lemma 3.22. If Assumption 3.21, equivalent to Assumption S1, holds, then there is a multiplier function M(s) such that

$$\text{(i) Re } M(j\omega - \beta_{sh}) \geq \delta_1 > 0 \quad ; \tag{3.2}$$

$$\text{(ii) Re } M(j\omega - \beta_{sh})G(j\omega - \beta_{sh}) \geq \delta_2 > 0 \tag{3.3}$$

where δ_1 and δ_2 are constants, for all $\omega \in (-\infty,\infty)$.

Proof. The Nyquist diagram assumption implies that the argument of $G(j\omega - \beta_{sh})$ lies in the interval $(-\pi\ ,\ \pi)$. If the conclusions of the lemma are to be fulfilled, $M(j\omega - \beta_{sh})$ and $M(j\omega - \beta_{sh})G(j\omega - \beta_{sh})$ must clearly have arguments in $(-\pi/2,\ \pi/2)$. Upon recalling that

$$\arg\left\{M(j\omega - \beta_{sh})G(j\omega - \beta_{sh})\right\} = \arg\left\{M(j\omega - \beta_{sh})\right\} + \arg\left\{G(j\omega - \beta_{sh})\right\}\ ,$$

$M(j\omega)$ could be constructed as follows :

$$\text{Define } \psi(\omega) = (\pi/2) - \arg\left\{G(j\omega)\right\} \text{ for } \arg\left\{G(j\omega)\right\} \geq 0$$
$$= (-\pi/2) - \arg\left\{G(j\omega)\right\} \text{ for } \arg\left\{G(j\omega)\right\} < 0 \tag{3.4}$$

Then $\arg\left\{M(j\omega)\right\}$ lies in the band formed by two functions (Fig.3.1) ψ_1 and ψ_2 which are defined as follows :

$$\psi_1(\omega) = \left[(-\pi/2) + \varepsilon,\ \psi(\omega) - \varepsilon\right] \quad \text{for } \arg\left\{G(j\omega)\right\} \geq 0,$$
$$\psi_2(\omega) = \left[\psi(\omega) + \varepsilon,\ (\pi/2) - \varepsilon\right] \quad \text{for } \arg\left\{G(j\omega)\right\} < 0, \tag{3.5}$$

for some constant $\varepsilon > 0$. The rest of the construction, i.e., determination of $M(j\omega)$ from its phase angle behaviour, follows standard techniques, one of which consists in using Guillemin's method [G 6,p 316] for the driving-point impedance, i.e., a positive real function. The impedance function so constructed is strictly positive real and rational. Hence the lemma is proved.

Remark 3.23. Note that in the statement of Lemma 3.22, no a priori restriction on the positive realness or otherwise of the multiplier function is imposed. When the multiplier function is, in fact, strictly positive real and rational, then the converse of Lemma 3.22 is true. This is a special case of Theorem 3.41 of the present chapter. An alternative proof of the converse of Lemma 3.22 is

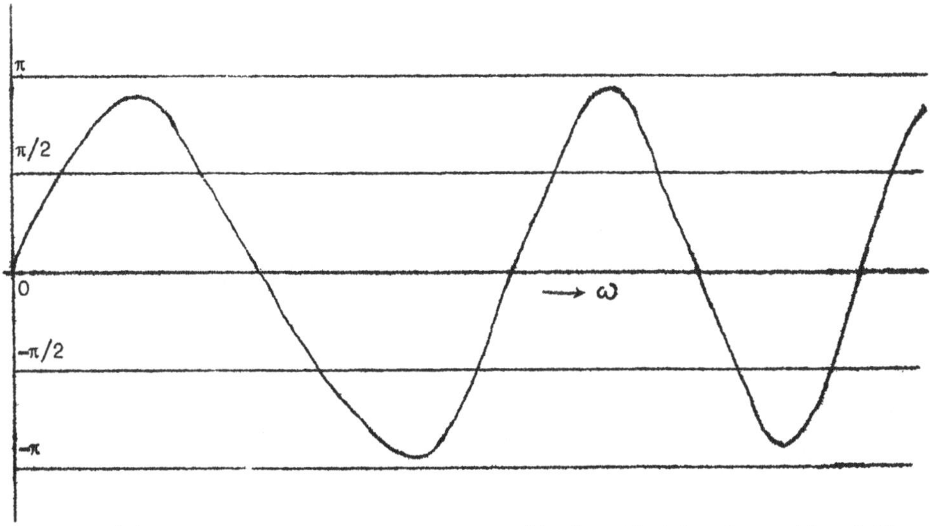

Fig.3.1(a) A typical phase angle plot of $G(j\omega)$ satisfying Assumption 3.21.

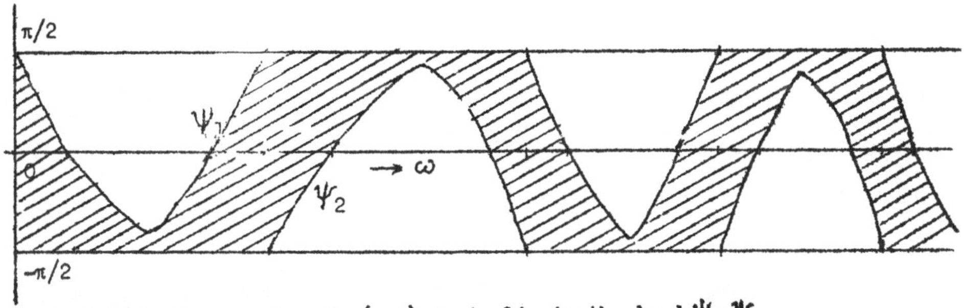

Fig.3.1(b) Phase angle of $M(j\omega)$ is to lie in the band ψ_1-ψ_2.

based on Lemma 3.21 concerning positive real functions, as found in [B 20, Theorem 2,p 257]. See Exercise 3.5. When the multiplier function is nonpositive real (and not necessarily rational), establishment of this converse in the Lyapunov framework of the chapter constitutes an open problem.

Remark 3.24. Intuitively speaking, stability of the linear time invariant system (2.15) depends on the amounts by which signals are amplified and delayed in flowing round the loop. Lemma 3.22 is the equivalent quantitative statement. The question then is : Are similar considerations involved in nonlinear time varying problems ? Naturally, the classical definitions of gain and phase shift, in terms of the frequency response have no strict meaning in nonlinear or time varying systems. However, based on Lemma 3.22 which gives an alternative characterization of the notion of phase shift, and on Exercise 2.3 which essentially involves the gain of the components in feedback loop, the basic problem of nonlinear time varying system stability can be formulated as follows :

Find a rational* multiplier function $M(s)$ with the arg $\{M(j\omega)\}$ in the band (3.5) such that (i) its introduction into the feedback system as in Exercise 2.3 does not lead to instability ;and (ii) the cascade connection of $M(s)$ and $k(t)\varphi(\cdot)$ gives rise to a 'phase shift', between its input and output, of less than $\pi/2$, i.e., $M(s)$ and $k(t)\varphi(\cdot)$ form a passive combination.

In the Lyapunov framework, this problem has been solved only partially : (i) The multiplier function used is positive real with a certain other constraint(s) imposed on it depending on the class of nonlinearities under consideration; (ii) The time varying gain $k(t)$ has to obey a certain upper or lower global bound on the normalized rate of variation, $(dk/dt/k(t))$, the bound depending, for the linear system, merely on the amount of damping in the time invariant system and, for the nonlinear system (2.1), also on the nature of the nonlinearity.

We employ positive real functions to generate quadratic functions of state

---

* This is because we are dealing with a finite dimensional system. For a more general statement , see Chapters 4 and 5 .

(for linear time varying system (2.21) stability analysis), and a quadratic function of state + integral of some function of the nonlinearity (for nonlinear time varying system (2.1) stability analysis). To this end, we employ the frequency domain information of $M(j\omega)$ for deriving positive definite or positive semidefinite quadratic forms and nonnegative integrals involving the nonlinearity of the system (2.1)

In what follows, $Z(s) = m(s)/n(s)$ where $m(s)$ and $n(s)$ are finite polynomials in s.

3.3. Generation of quadratic forms [B 16(b)] .

(a) Suppose $u(t)$ is n times differentiable with respect to t and that α_{ij} are constants. Then the integral

$$I = \int_{t_1}^{t_2} \sum_{i=1}^{n} \sum_{j=1}^{n} \alpha_{ij}(d^i u/dt^i)(d^j u/dt^j)dt \tag{3.6}$$

is defined and depends not only on the value taken by u and its derivatives at t_1 and t_2 but also on their values between t_1 and t_2. However, if the integrand in (3.6) is of the form $(u(t)(du/dt))$, then $I = (u^2(t_2) - u^2(t_1))/2$, independent of the form of u. By extension, the integral defined by (3.6) is said to be independent of the path if it can be evaluated in terms of u, (du/dt), $\cdots$, $(d^{n-1}u/dt^{n-1})$ at t_2 and t_1. Observe that I is then a quadratic form in u and its first (n-1) derivatives.

Following Brockett [B 16(b)] , the notation

$$I = \int_{t(\underline{0})}^{t(\underline{x})} \sum_{i=1}^{n} \sum_{j=1}^{n} \alpha_{ij}(d^i u/dt^i)(d^j u/dt^j)dt = \underline{x}' P \underline{x} \tag{3.7}$$

describes a path integral (when independent of the path) which starts at $\underline{0}$ and ends at the point $\underline{x}$. Thus the integral is to be treated as a line integral in the state space. It should be noted that (3.7) is essentially the same as the one used by Malkin quoted in [B 9] :

$$I_m = \int_{t}^{\infty} \underline{x}'(\exp A(\tau - t))' C(\exp A(\tau - t)) \underline{x}\, d\tau \tag{3.8}$$

where the constant matrix is positive definite. Just as the time-derivative of

(3.8) is simply $(-\underline{x}'C\,\underline{x})$, so can the time-derivative of (3.7) be obtained by removing the integral sign :

$$\frac{dI}{dt} = \sum_{i=1}^{n} \sum_{j=1}^{n} \alpha_{ij}(d^i x/dt^i)(d^j x/dt^j) = (d\,\underline{x}'/dt)\;Q(d\underline{x}/dt) \qquad (3.9)$$

where Q is a constant matrix.

The value of this derivative along, for instance, the solutions of (2.4) (and hence of (2.1)) is obtained by replacing $(d\underline{x}/dt)$ in (3.9) by the right hand side of (2.4). Such an evaluation is denoted by

$$\left.\frac{dI}{dt}\right|_{(2.4)} = (\underline{x}'A_o' - k(t)\;\underline{b}'\varphi(\underline{c}'\underline{x}))\;Q(A_o\underline{x} - k(t)\;\underline{b}\;\varphi(\underline{c}'\underline{x})).$$

3.3(b). If $u(s)$ is a polynomial with real coefficients, let $\mathrm{Ev}\;u(s) = (u(s) + u(-s))/2$ denotes its even part. Note that for real ω, $\mathrm{Ev}\;u(j\omega) = \mathrm{Re}\;u(j\omega)$. If $u(s)$ is even and $\mathrm{Re}\;u(j\omega) \geq 0$ for all real ω, then according to a theorem of Wiener, there exists a unique polynomial $v(s)$ with real positive coefficients such that $v(s)v(-s) = u(s)$, and $v(s)$ has zeros only in the closed left half plane $\mathrm{Re}\;s \leq 0$. Let $(u(s))^{(+)}$ stand for $v(s)$ and $(u(s))^{(-)}$ for $v(-s)$. The latter is also known as the right half plane spectral factor of the even polynomial $u(s)$. Observe that $(u(s))^{(-)}$ has zeros only in the closed right half plane $\mathrm{Re}\;s \geq 0$. If $u_1(s)$ and $u_2(s)$ are two polynomials with real coefficients and $\mathrm{Re}\;(u_1(j\omega)/u_2(j\omega)) \geq 0$ for all real ω, then clearly $\mathrm{Re}(u_1(j\omega)u_2(-j\omega)) \geq 0$ for all real ω, and there is a unique factor $(\mathrm{Ev}\;u_1(s)u_2(-s))^{(-)}$.

As we propose to generate quadratic functions from the positive realness of functions $Z(s)$ and $Z(s)G(s)$, we need to apply the above results to $Z(j\omega)$ and $Z(j\omega)G(j\omega)$. We accomplish this below.

Given that $\mathrm{Re}\;Z(j\omega - \alpha) \geq 0$, $\mathrm{Re}\;Z(j\omega - \beta)G(j\omega - \beta)) \geq 0$ for all real ω and for some constants α and β, then the polynomial functions $r_1(s) = (\mathrm{Ev}\;m(s-\alpha)n(-s-\alpha))^{(-)}$ and $r_2(s) = (\mathrm{Ev}\;m(s-\beta)q(s-\beta)n(-s-\beta)\,p(-s-\beta))^{(-)}$ are uniquely defined. The polynomial operators $r_1(D)$ and $r_2(D)$ appear in the integrands of the integrals of the type (3.7).

3.3 (c). In order to motivate the statement of the next lemma concerning the generation of quadratic forms, we consider the function $Z(s) = (\gamma_1 s + \gamma_o)$ for some constants γ_1 and γ_o. Here $m(s) = (\gamma_1 s + \gamma_o)$ and $n(s) = 1$. Note that $\text{Re } Z(j\omega) \geq 0$ if $\gamma_o \geq 0$. Then $(\text{Ev } m(s)n(-s))^{(-)} = \sqrt{\gamma_o}$. The integral

$$I_1 = \int_{t(x^1)}^{t(x^2)} \left\{ (\gamma_1(dw/dt) + \gamma_o w)w - \gamma_o w^2 \right\} dt \tag{3.10}$$

is independent of the path, and the last term of the integrand can be written as $((\text{Ev } m(D)n(-D))^{(-)}w)^2$.

If $Z(s)$ is positive real, we require that γ_1 and γ_o be nonnegative and at least one of them is positive. Alternatively, from Lemma 3.21, $\gamma_o \geq 0$ and the root of the equation $\gamma_1 s + \gamma_o + 1 = 0$ should be in the open left half of the complex plane. That is $\gamma_1 > 0$. (The case $\gamma_1 = 0$ is trivial). Now consider the differential equation

$$\gamma_1(dx/dt) + (\gamma_o + 1)x = 0 \tag{3.11}$$

which has solutions tending to zero for any arbitrary initial condition $x(0) = x_o$. That is, with $t(x_o) = 0$, $t(0) = \infty$, integration of (3.10) along the solutions of (3.11) with $x^1 = 0$ and $x^2 = x_o$ implies that w be replaced by x in the integrand of (3.10) :

$$I_1 \Big|_{(3.11)} = \int_{\infty}^{0} -(1+\gamma_o)x^2\,dt = \int_0^{\infty} (1+\gamma_o)x^2\,dt$$

which is zero only for $x = 0$. Noting that the order of (3.11) is one, I_1 is positive definite.

Remark 3.31. Equation (3.10) is defined and is independent of path for $\gamma_1 < 0$ also, i.e., for $Z(s)$ nonpositive real but with $\text{Re } Z(j\omega) \geq 0$ for all ω. In this case, (3.11) has solutions tending to ∞ for an arbitrary initial condition $x(0) = x_o \neq 0$. Now choose $t(x_o) = 0$ and $t(0) = -\infty$ and set $x^1 = 0$ and $x^2 = x_o$. Then (3.10) along the solutions of (3.11) becomes

$$I_1 \Big|_{(3.11)} = -\int_{-\infty}^{0} (\gamma_o + 1)x^2\,dt$$

which is negative definite. However, this interpretation has not been employed in literature, and seems to be an interesting possibility which could answer the question posed under Remark 3.23 above.

We now state a lemma which generalizes what has been derived above and constitutes an important result.

Lemma 3.31. The path integral defined by

$$I_2 = \int_{t(\underline{w}^1)}^{t(\underline{w}^2)} \left\{ m(D)w\ n(D)w - ((Ev\ m(D)n(-D))^{(-)}w)^2 \right\} d\tau \tag{3.12}$$

evaluated along the trajectories (or solutions) of

$$m(D)w + n(D)w = 0 \tag{3.13}$$

is (i) path independent if $\operatorname{Re}(m(j\omega)/n(j\omega)) \geq 0$ for all real ω ; and (ii) a positive definite quadratic form in $\underline{w}$, the state space of (3.13), if $m(s)/n(s)$ is positive real and there are no commom factors between $m(s)$ (or $n(s)$) and $(Ev\ m(s)n(-s))^{(-)}$.

Proof. The first part can be established by an extension of what has been said above for the integral I_1 defined by (3.10). For this, $m(D)$ and $n(D)$ are to be written in polynomial form, and the integral (3.12) integrated by parts.

Evaluation of (3.12) along the solutions of (3.13) gives

$$I_2 \Big|_{(3.13)} = \int_0^{\infty} \left\{ (n(D)w)^2 + ((Ev\ m(D)n(-D))^{(-)}w)^2 \right\} d\tau$$

or

$$I_2 \Big|_{(3.13)} = \int_0^{\infty} \left\{ (m(D)w)^2 + ((Ev\ m(D)n(-D))^{(-)}w)^2 \right\} d\tau$$

from which the second part of the lemma follows.

Remark 3.32. Suppose that in Lemma 3.31, $m(s)$ is a polynomial of degree η. Then the degree of $n(s)$, by virtue of the positive realness of $Z(s) \triangleq m(s)/n(s)$, is η, $\eta-1$ or $\eta+1$. Hence order of the system (3.13) is η or $\eta+1$, and the dimension of the state space of (3.13) is η or $\eta+1$. Further note that the common factors between $m(s)$ or $n(s)$ and $(Ev\ m(s)n(-s))^{(-)}$ need be searched only on the imaginary axis. This is because $m(s)$ (or $n(s)$), in view of the positive realness of $Z(s)$, has zeros only in the closed left half plane and $(Ev\ m(s)n(-s))^{(-)}$

has zeros only in the closed right half plane.

Corollary 3.311. The path integral

$$I_3 = \int_{t(\underline{w}^1)}^{t(\underline{w}^2)} \left\{ p(D)n(D)w\ q(D)m(D)w - ((Ev\ q(D)m(D)p(-D)n(-D))^{(-)}w)^2 \right\} d\tau \qquad (3.14)$$

integrated along the solutions of

$$p(D)n(D)w + q(D)m(D)w = 0 \qquad (3.15)$$

is (i) path independent if $Re(q(j\omega)m(j\omega)/p(j\omega)n(j\omega)) \geq 0$ for all real ω and (ii) positive definite in $\underline{w}$, the $(\eta + n)$-dimensional state space of (3.15), if Z s)G(s) is positive real and there are no common factors on the imaginary axis between $p(s)n(s)$ (or $q(s)m(s)$) and $(Ev\ q(s)m(s)p(-s)n(-s))^{(-)}$.

Corollary 3.312. In Lemma 3.31, replace w by $(\exp(\alpha t))q(D)w$, D by $D-\alpha$, s by $(s-\alpha)$ for some $\alpha \geq 0$ and integral I_2 by $I_2' = \exp(-2\alpha t)I_2$. Then I_2' is a positive semi-definite quadratic form in an enlarged state space of dimension at the most, $\eta + n$. In Corollary 3.311, replace w by $(\exp(\beta t))w$, D by $D-\beta$, s by $(s-\beta)$ for some $\beta \geq 0$, and integral I_3 by $I_3' = \exp(-2\beta t)I_3$. Then I_3' is a positive definite quadratic form in the $\eta + n$ dimensional state space of (3.15).

The next step in the stability analysis, as explained in Sec.2.52, is the use of I_2' and I_3' in suitable combination as a Lyapunov function candidate and evaluation of its time derivative along the system (2.1) solutions. However, (i) the dimension of the state space in which the quadratic forms I_2' and I_3' are defined is different from the dimension of the state space of (2.1) ; and (ii) it turns out that for the linear system (2.21), a particular case of (2.1), I_2' and I_3' suffice, whereas for the nonlinear system (2.1), in the place of I_2', a new integral involving the nonlinearity is to be defined. As regards the former, we consider an auxiliary system in whose state space we imbed the state space of (2.1) or (2.4), the relation between the auxiliary system and (2.1) or (2.4) being given by the following lemma. The integral involving the nonlinearity will be taken up later.

Lemma 3.32. If the system

$$p(D)n(D)y + k(t)\varphi(q(D)n(D)y) = 0 \tag{3.16}$$

is exponentially stable and $n(D)w = 0$ represents an asymptotically stable system, then the system (2.1) is also exponentially stable.

Proof. If $\phi(t)$ is a solution of (3.16), then $n(D)\phi(t)$ is a solution of (2.1). If all the solutions of (3.16) tend to zero exponentially, so do $\phi(t)$ and its first $(\eta + n - 1)$ derivatives. Denoting $n(D)\phi(t)$ by $\phi'(t)$, we conclude that $\phi'(t)$ and its first $(n - 1)$ derivatives decay to zero exponentially. This completes the proof.

Corollary 3.321. If the system

$$P(D)n(D)y + k(t)q(D)n(D)y = 0 \tag{3.17}$$

is exponentially stable and $n(D)w = 0$ represents an asymptotically stable system, then the linear system (2.21) is also exponentially stable.

3.3(c). The basic stability result of Corduneanu [C 8].

By way of preliminaries, we now state a lemma, an extension (due to Corduneanu [C 8]) of Lyapunov's [L 10] stability theorem, which forms the main tool in establishing exponential stability criteria for (2.1) and (2.21).

Lemma 3.33. If there exist a quadratic form $v(\underline{x},t) \triangleq \underline{x}'P(t)\underline{x}$ having the property

$$\gamma_o \underline{x}'\underline{x} \leq v(\underline{x},t) \leq \gamma_1 \underline{x}'\underline{x} \tag{3.18}$$

for all $t \geq t_o \geq 0$ for some positive constants γ_o and γ_1 with $\gamma_o < \gamma_1$, and real valued function $\lambda(t)$ on $[t_o, \infty)$ such that the derivative of $v(\underline{x},t)$ along the solutions of (2.4) (and hence of (2.1)) satisfies the inequality

$$\left.\frac{dv}{dt}\right|_{(2.4)} \leq \lambda(t)v \tag{3.19}$$

then there exists a positive constant η_o such that the solutions of (2.4) satisfy the inequality

$$\|\underline{x}(t)\| \leq \eta_o \|\underline{x}(t_o)\| \exp(1/2 \int_{t_o}^{t} \lambda(\tau)d\tau). \tag{3.20}$$

Proof. Integration of the differential inequality (3.19) yields

$$v(\underline{x},t) \leq v(\underline{x}_o,t_o)\exp\left(\int_{t_o}^{t} \lambda(\tau)d\tau\right). \tag{3.21}$$

Consequently, from (3.18) and (3.21), we have

$$\gamma_o\|\underline{x}\|^2 \leq v(\underline{x},t) \leq \gamma_1\|\underline{x}_o\|^2 \exp\left(\int_{t_o}^{t} \lambda(\tau)d\tau\right)$$

and hence

$$\|\underline{x}\|^2 \leq (\gamma_1/\gamma_o)\|\underline{x}_o\|^2 \exp\left(\int_{t_o}^{t} \lambda(\tau)d\tau\right)$$

from which follows the inequality (3.20) with $\eta_o = (\gamma_1/\gamma_o)^{1/2}$. The lemma is proved.

Corollary 3.331. If, in Lemma 3.33, $v(\underline{x},t)$ is nonquadratic and has the property

$$\gamma_o\underline{x}'\underline{x} \leq v(\underline{x},t) \leq \gamma_1\underline{x}'\underline{x} + \gamma_2\exp(-\beta t)\int_{t_o}^{t} \exp(\beta\tau)\underline{x}'(\tau)\underline{x}(\tau)d\tau \tag{3.22}$$

for all $t \geq t_o \geq 0$, some positive constants γ_o and γ_1 with $\gamma_o < \gamma_1$, nonnegative constants γ_2 and β, then the conclusion (3.20) of the lemma holds for the system (2.1). No alteration in its proof is warranted, because $v(\underline{x}_o,t_o)$ in (3.21) still satisfies the inequality $v(\underline{x}_o,t_o) \leq \gamma_1\|\underline{x}_o\|^2$.

Corollary 3.332. If, in Lemma 3.33, $\lambda(\cdot)$ satisfies the integral inequality

$$(1/T)\int_{t_o}^{t_o+T} \lambda(\tau)d\tau \leq -\nu$$

for all $T > 0$ and some positive constant ν, then

$$\|\underline{x}(t)\| \leq \eta_o\|\underline{x}_o\|\exp(-\nu(t-t_o)/2)$$

and the system (2.4) (and hence (2.1)) is exponentially stable.

Remark 3.33. Lemma 3.33, as applied to the auxiliary system (3.16), requires that $\underline{x}$ be replaced by the state vector for (3.16). Note also that the lemma reduces to Lyapunov's classic stability theorem [L 10] if $\lambda(t) = -\eta_1$ for all $t \geq t_o \geq 0$ and some positive constant η_1.

Using the above preliminaries, we derive below (Sec.3.4) exponential stability criteria for the linear system (2.21). The nonlinear system (2.1) will be considered in Sec.3.5.

3.4. Linear System Exponential Stability.

(a) Let the state vector of (3.17) be denoted by $\underline{\chi}(t)$ of which the state vector $\underline{x}(t)$ of (2.21) is a subspace. Assuming that (i) $Z(s-\alpha)$ and $Z(s-\beta)G(s-\beta)$ are positive real for some constants $\alpha \geq 0$ and $\beta > 0$, (ii) there are no common factors on the imaginary axis between $p(s-\beta)n(s-\beta)$ (or $q(s-\beta)m(s-\beta)$) and $(\mathrm{Ev}\ q(s-\beta)m(s-\beta)p(-s-\beta)n(-s-\beta))^{(-)}$, we can generate the following quadratic forms by suitably choosing the path of integration as indicated in Lemmas 3.31 and 3.32. The change we now make is the replacement of the $\eta + n$ dimensional vector in Corollary 3.312 by $\underline{\chi} = \mathrm{col}[\chi_1, \chi_2, \cdots, \chi_{\eta+n}]$ and of w inside the integrals by $\chi \triangleq \chi_1$. This does not affect the properties of the quadratic forms generated by means of path integrals.

For convenience, define

$$r_1(s) = (\mathrm{Ev}\ q(s-\beta)m(s-\beta)p(-s-\beta)n(-s-\beta))^{(-)},$$
$$r_2(s) = (\mathrm{Ev}\ m(s-\alpha)n(-s-\alpha))^{(-)} .$$

Then, based on Corollaries 3.311 and 3.312, we have the following quadratic forms in $\underline{\chi}$:

$$V_1(\underline{\chi},t) = \exp(-2\beta t)\int_{t(\underline{0})}^{t(\underline{\chi})}\Big\{p(D-\beta)n(D-\beta)(\chi\exp(\beta\tau))q(D-\beta)m(D-\beta)(\chi\exp(\beta\tau)) - (r_1(D)(\chi\exp(\beta\tau)))^2\Big\}\,d\tau \tag{3.23}$$

$$V_2(\underline{\chi},t) = \exp(-2\alpha t)\int_{t(\underline{0})}^{t(\underline{\chi})}\Big\{m(D-\alpha)(\exp(\alpha\tau)q(D)\chi)n(D-\alpha)(\exp(\alpha\tau)q(D)\chi) - (r_2(D)\ (\exp(\alpha\tau)q(D)\chi))^2\Big\}\,d\tau \tag{3.24}$$

Note that $V_1(\underline{\chi},t)$ has been made positive definite and $V_2(\underline{\chi},t)$ positive semidefinite by choosing suitable paths for integration. The time-derivatives of $V_1(\underline{\chi},t)$ and $V_2(\underline{\chi},t)$ are given by

$$(dV_1(\underline{\chi},t)/dt) = -2\beta\ V_1(\underline{\chi},t) + (p(D)n(D)\chi)(q(D)m(D)\chi) - (r_1(D+\beta)\chi)^2, \tag{3.25}$$

$$(dV_2(\underline{\chi},t)/dt) = -2\alpha\ V_2(\underline{\chi},t) + (m(D)q(D)\chi)(n(D)q(D)\chi) - (r_2(D+\alpha)\chi)^2 \tag{3.26}$$

Next, we evaluate these derivatives along the solutions of (3.17), the system whose stability is to be investigated. This amounts to replacing $\underline{\chi}$ and χ in (3.25) and (3.26) by the solution vector $\underline{\chi}(t)$ of (3.17) and y respectively. Hence we have

$$(dV_1(\underline{\chi},t)/dt)\Big|_{(3.17)} = -2\beta\ V_1(\underline{\chi}(t),t)-k(t)(m(D)q(D)y)(n(D)q(D)y)-(r_1(D+\beta)y)^2, \quad (3.27)$$

$$(dV_2(\underline{\chi},t)/dt)\Big|_{(3.17)} = -2\alpha\ V_2(\underline{\chi}(t),t)+(m(D)q(D)y)(n(D)q(D)y)-(r_2(D+\alpha)y)^2 \quad (3.28)$$

Let
$$V(\underline{\chi},t) \triangleq V_1(\underline{\chi},t) + k(t)V_2(\underline{\chi},t)\ , \quad (3.29)$$
and
$$\theta(t) \triangleq (dk/dt/k). \quad (3.30)$$

It is easy to verify that

$$(V(\underline{\chi},t)/dt)\Big|_{(3.17)} = -2\beta\ V_1(\underline{\chi}(t),t)+(\theta(t)-2\alpha)k(t)V_2(\underline{\chi}(t),t)-(r_1(D+\beta)y)^2$$
$$-k(t)(r_2(D+\alpha)q(D)y)^2 \quad (3.31)$$

In order to write the expression for the time-derivative of $V(\underline{\chi},t)$ along the solutions of (3.17) in the form of inequality (3.19), we need the following lemma whose proof is obvious.

Lemma 3.41. Let $\lambda_1(t)$ and $\lambda_2(t)$ be bounded real functions on $[0,\infty)$. If $V_1(\underline{\chi},t)$ and $V_2(\underline{\chi},t)$ are nonnegative for all t in $[0,\infty)$, and $V_{12}(\underline{\chi},t) = \lambda_1(t)\ V_1(\underline{\chi},t) + \lambda_2(t)\ V_2(\underline{\chi},t)$, then

$$V_{12}(\underline{\chi},t) \leq \sup_{t\geq 0} [\lambda_1(t), \lambda_2(t)]\ (V_1+ V_2)(\underline{\chi},t).$$

Based on Lemma 3.41, we have from (3.31)

$$(dV(\underline{\chi},t)/dt)\Big|_{(3.17)} \leq \sup_{t\geq 0} (-2\beta,\theta(t)-2\alpha)V(\underline{\chi}(t),t) \quad (3.32)$$

Let $\lambda(t) = \sup_{t\geq 0} (-2\beta,\theta(t)-2\alpha)$. Then (3.32) has the form of inequality (3.19). Further, the quadratic form (3.30) satisfies the hypothesis of Lemma 3.33 with $\underline{x}$ replaced by $\underline{\chi}$. As a first step towards a general exponential stability condition for (2.21), we have the following theorem..

Theorem 3.41. If (i) Z(s) (the 'multiplier') is a function of the complex variable s, such that $Z(s-\alpha)$ is positive real for some constant $\alpha \geq 0$;

(ii) $Z(s-\beta)G(s-\beta)$ is positive real for some constant $\beta > 0$, and $m(s-\beta)q(s-\beta)$ (or $n(s-\beta)p(s-\beta)$) and $(\mathrm{Ev}\ q(s-\beta)m(s-\beta)p(-s-\beta)n(-s-\beta))^{(-)}$ have no common factors on the imaginary axis; and with $(\theta(t)+2(\beta-\alpha))^+$ denoting the positive lobes of $(\theta(t)+2(\beta-\alpha))$,

(iii) $$(1/T)\int_{t_o}^{t_o+T} (\theta(t)+2(\beta-\alpha))^+ dt \leq 2\beta-\nu \qquad (3.33)$$

for all $T > 0$, for some positive constant ν , then the system (2.21) is exponentially stable.

Proof. As a Lyapunov-Corduneanu function candidate for (3.17) and hence for (2.21), choose $V(\underline{\chi},t)$ defined by (3.29) with $V_1(\underline{\chi},t)$ and $V_2(\underline{\chi},t)$ given by (3.23) and (3.24) respectively. $V(\underline{\chi},t)$ satisfies the hypotheses of Lemma 3.34. Its time derivative along the solutions of (3.17) satisfies the inequality (3.32). Invoke Lemma 3.33 (Corollary 3.332) to conclude that hypothesis (iii) implies exponential stability of (3.17) and hence, by virtue of Lemma 3.32 (Corollary 3.321), of (2.21).

Remark 3.41. Hypothesis (iii) of Theorem 3.41 can be improved to the following form :

(iv) $$(1/T)\int_{t_o}^{t_o+T} (\theta(t)+2(\beta-\alpha))^+ dt \leq M < \infty,$$ and for all finite $T > 0$ and for some positive constant M ; (3.34)

$$\lim_{T\to\infty} (1/T)\int_{t_o}^{t_o+T} (\theta(t)+2(\beta-\alpha))^+ dt \leq 2\beta-\nu$$ for some positive constant ν .

There are two methods of accomplishing this improvement.

Method 1. Let $\zeta(t)$ be a nonnegative (integrable and bounded) function on $[t_o,\infty)$, and $h_o(t) = \exp(-\int_{t_o}^{t}\zeta(\tau)d\tau)$. Assume that the integral $\int_{t_o}^{t}\zeta(\tau)d\tau \leq M < \infty$ for all t in $[t_o,\infty)$ for some positive constant M, and

$$0 < \varepsilon \leq \lim_{t\to\infty}\int_{t_o}^{t} \zeta(\tau)d\tau \leq M < \infty.$$

Then $h_o(t)$ is a bounded positive function. Note that

$$\left(\frac{dh(t)}{dt} / h(t) \right) = -\zeta(t)$$

which is nonpositive. Let

$$V_o(\underline{\chi},t) = h_o(t)\left\{V_1(\underline{\chi},t)+k(t)V_2(\underline{\chi},t)\right\}$$

where $V_1(\underline{\chi},t)$ and $V_2(\underline{\chi},t)$ are defined by (3.23) and (3.24) respectively. It can be shown using $V_o(\underline{\chi},t)$ that Theorem 3.41 holds with hypothesis (iii) replaced by hypothesis (iv), represented by the set of inequalities (3.34).

Method 2. This consists in a modification of the above method, and seems to be more versatile, as it will be used in Chapters 4 and 5 to derive improved versions of the global bound (3.34) on $\theta(t)$. To describe the method, we need the following definition.

Definition 3.41. Let $\mathcal{K}$ be the class of absolutely continuous real valued functions $\underline{k}(\cdot)$ on $[0,\infty)$ with each $k(\cdot)$ having constants $\underline{k} > 0$ and $\bar{k} \geq \underline{k}$ for which $\underline{k} \leq k(t) \leq \bar{k}$ for all $t \geq 0$.

Let $h(t) \in \mathcal{K}$. Define

$$V_o'(\underline{\chi},t) = h(t)\left\{V_1(\underline{\chi},t)+k(t)V_2(\underline{\chi},t)\right\} \tag{3.35}$$

where $V_1(\underline{\chi},t)$ and $V_2(\underline{\chi},t)$ are given by (3.23) and (3.24) respectively. Then

$$\left.\frac{dV_o'(\underline{\chi},t)}{dt}\right|_{(3.17)} = \left(\left(\frac{dh}{dt}/h\right)-2\beta\right)h(t)V_1(\underline{\chi}(t),t)$$
$$+\left(\left(\frac{dh}{dt}/h\right)+\theta(t)-2\alpha\right)h(t)k(t)V_2(\underline{\chi}(t),t)$$
$$-h(t)\left\{(r_1(D+\beta)y)^2+k(t)(r_2(D+\alpha)y)^2\right\} \tag{3.36}$$

We now state a lemma which serves to enlarge the stability boundary obtained from (3.36) by using the classic Lyapunov theorem. (See Remark 3.33 above).

Lemma 3.42. Suppose for functions $h(\cdot)$ and $k(\cdot)$ belonging to $\mathcal{K}$, the following differential inequalities hold :

$$\left(\frac{dh}{dt}/h\right)-2\beta \leq -\nu_1, \tag{3.37}$$

$$\left(\frac{dk}{dt}/k\right)+\left(\frac{dh}{dt}/k\right)-2\alpha \leq -\nu_2 \tag{3.38}$$

for some positive constants ν_1 and ν_2. Further, suppose hypotheses (i) and (ii) of Theorem 3.41 are satisfied. Then (i) the system (3.17) and hence (2.21) is exponentially stable ; and (ii) the set of integral inequalities (3.34) is satisfied.

Proof. The first part follows from the classic Lyapunov theorem. For proving the second part, observe that choice of

$$\left(\frac{dh}{dt}/h\right) = -(\theta(t)+2(\beta-\alpha))^{+}+2\beta-\nu \tag{3.39}$$

for some constant $\nu > 0$ satisfies both (3.38) and (3.37). Integrate (3.39) to get

$$h(t) = h(t_o)\exp\left(-\int_{t_o}^{t} (\theta(\tau)+2(\beta-\alpha))^{+}d\tau +(2\beta-\nu)(t-t_o)\right). \tag{3.40}$$

Since $h(t)\in\mathcal{K}$,

$$-M_1 \leq -\int_{t_o}^{t} (\theta(\tau)+2(\beta-\alpha))^{+}d\tau+(2\beta-\nu)(t-t_o) \leq M_2 \tag{3.41}$$

for all $t \geq t_o$, for some positive constants M_1 and M_2. The integral inequality (3.41) can be rewritten as a combination of two inequalities :

$$\int_{t_o}^{t} (\theta(\tau)+2(\beta-\alpha))^{+}d\tau \leq (2\beta-\nu)(t-t_o)+M_1 \tag{3.42}$$

and

$$\int_{t_o}^{t} (\theta(\tau)+2(\beta-\alpha))^{+}d\tau \geq -(2\beta-\nu)(t-t_o)-M_2 \tag{3.43}$$

for all $t \geq t_o$, for some positive constants M_1 and M_2. Inequality (3.43) is trivially satisfied. Hence, only the inequality (3.42) is of relevance. This can be reduced to

$$(1/T)\int_{t_o}^{t_o+T} (\theta(\tau)+2(\beta-\alpha))^{+}d\tau \leq (2\beta-\nu)+(M_1/T) \tag{3.44}$$

for all $T > 0$. The set of inequalities (3.34) is equivalent to (3.44). The lemma is proved.

Remark 3.42. Note that, in hypotheses, (i) and (ii) of Theorem 3.41, $\alpha \geq \beta$. If we were to set $\alpha = \beta$, we would be to some extent weakening the global condition on $\theta(t)$ as evident from (3.34). However, it is then possible to invoke Lemma 3.22 to express hypotheses (i) and (ii) of Theorem 3.41 in geometric terms. This restatement in combination with the set of inequalities (3.34) yields the following theorem which is a geometric version of Theorem 3.41.

Theorem 3.42. Let $\beta_{sh} > 0$ be the maximal shift for which the plot of $G(j\omega - \beta_{sh})$ avoids the negative real axis with ω in $(-\infty, \infty)$. Then the linear system (2.21) is exponentially stable if $\theta(t)$, the normalized rate of variation of $k(t)$, satisfies the global constraint

$$(1/T) \int_{t_o}^{t_o+T} \theta^+(t)dt \leq M < \infty$$

for all finite $T > 0$ and for some positive constant M; and

$$\lim_{T \to \infty} (1/T) \int_{t_o}^{t_o+T} \theta^+(t)dt \leq 2\beta_{sh} - \nu$$

for some positive constant ν.

See Exercise 3.6 for a method of deriving an equivalent exponential stability bound in terms of $\theta^-(t)$.

3.4(b). Comparison with other results in literature.

i) The local or point-by-point condition

$$(dk/dt) \leq 2\alpha k(t) \quad \text{for all } t \geq 0$$

obtained by Gruber and Willems [G 5] using Lyapunov's theorem (see Remark 3.33) is a special case of the global bound (3.33).

ii) Both the global bounds (3.33) and (3.34) involve only the positive lobes of $\theta(t)+2(\beta-\alpha)$. The negative lobes of $\theta(t)+2(\beta-\alpha)$, which can be arbitrarily large, are immaterial. Compare this with the bound

$$\sup_{t \geq 0} (1/T) \int_{t}^{t+T} \left| d \log k(\tau)/d\tau \right| d\tau < 4\beta_{sh} \tag{3.46}$$

obtained by Freedman and Zames [F 6] on the basis of the inequalities (3.37) and (3.38) involving $h(t) \in \mathcal{K}$. Note that the integrand in (3.46) is just $\theta(\tau)$ and hence both the positive and negative lobes of $\theta(t)$ contribute to the value of the integral. This is in contrast with (3.45) where only the positive lobes of $\theta(t)$ are to be considered. In fact, the method of deriving the integral bound (3.46) in [F 6] has been improved by Acker [A 1] leading to a result of the type (3.45).

iii) The earlier results of Brockett and Forys [B 17] and others are special cases of Theorem 3.41. The multiplier function employed in [B 17] is an RC-RL impedance function which is a subclass of the general positive real functions. The geometric interpretation of the stability criterion of [B 17] as attempted by Freedman [F 5(b)] is not wholly correct because of the explicit statement and use of the constraint [F 5(b), pp 563-564]

$$\lim_{|\omega|\to\infty} \arg \left\{ G(j\omega) + (1/K) \right\} = 0 \tag{3.47}$$

where we now have K arbitrarily large and hence (1/K) can be removed from the expression. Even the simplest second order transfer function $(1/(s+\alpha_1)(s+\alpha_2))$ for positive constants α_1 and α_2 does not obey (3.47).

Other criteria like those of Kuh [K 8] involve the evaluation of the minimum and maximum eigenvalues of the matrix $(A_0-k(t)\underline{b}\,\underline{c}')$ which is quite complicated. See Exercise 3.7.

iv) One of the stability criteria of Davis [D 3(a)] involves the maximum of the average square of k(t) over an interval of length T :

With $\langle k\rangle^2 \triangleq \sup_{t\geq 0} (1/T)\int_{t-T}^{T} |k(\tau)|^2 d\tau$, suppose that

$$\langle k\rangle^2 \sup_{\omega\in R} \left\{ |q(j\omega)/p(j\omega)|^2 \left((\omega T/3)^2 + 1\right) \right\} < (1/2),$$

then the feedback system (2.21) is L_2-stable (see Definition 2.35) provided that each zero of the polynomial p(s) has a negative real part. As indicated in Chapter 2 (Sec.2.41), exponential stability of (2.21) is equivalent to its L_2-stability [A 4] . The criteria of Davis [D 3(a)] which are in fact derived from the circle criterion have the following drawbacks : (1) Even the circle criterion cannot be obtained as a special case. (2) It is not clear how to choose the averaging interval T to 'best advantage in each specific case.' For instance, how can one choose T for the Mathieu equation ? See also Exercise 3.8.

Remark 3.43. If hypothesis (ii) of Theorem 3.41 is satisfied for Z(s) = 1, then evidently, hypothesis (iii) can be bypassed. That is, no restriction is placed on the rate of variation of k(t). The theorem is then equivalent to the circle criterion. (See Exercise 3.4). If k(t) is constant for all $t \geq 0$, then Theorem

3.41 reduces to the Nyquist criterion for linear time invariant systems.

Remark 3.44. Theorems 3.41 and 3.42 appear to be successful only to the extent that $\beta_{sh} > 0$. If $\beta_{sh} = 0$, no exponential stability conditions can be derived. Even with $\beta_{sh} > 0$, the stability boundary obtained from Theorem 3.41 for a lightly damped Mathieu equation (Exercise 3.8) is no match for the classical boundary as found in McLachlan [M 7] and Hayashi [H 6]. Apart from this, the additional information of periodicity of the time varying gain in the Mathieu equation cannot be utilized to advantage while deriving the stability criteria. This is in contrast with the Popov and Zames - Sandberg methods of Chapters 4 and 5 respectively.

3.5. Nonlinear System Exponential Stability.

(a) Preliminaries. As explained above in Sec. 3.3, we need to use an integral containing the nonlinear function $\varphi(\cdot)$ along with $V_1(\underline{X}, t)$ defined by (3.23) as a Lyapunov - Corduneanu function candidate for nonlinear system (2.1). We establish conditions for the exponential stability of the auxiliary system (3.16) in whose state space the state space of (2.1) is imbedded. See Lemma 3.32. In brief, we are looking for conditions under which a driving point impedance function in cascade with the nonlinearity of (2.1) gives rise to a 'passive' system i.e., input multiplied by the output of the combination integrated over time is always non-negative for all permissible input signals . As is to be expected, the characteristics of the nonlinearity play a major role in the stability conditions. Let

$$\Phi(\sigma) = \int_0^\sigma \varphi(w)dw \; ; \quad \delta(\sigma) = (\Phi(\sigma)/\varphi(\sigma)\sigma) \text{ for } \sigma \neq 0. \tag{3.48}$$

$$\delta_s = \sup_\sigma \delta(\sigma) \; ; \quad \delta_i = \inf_\sigma \delta(\sigma) \; . \tag{3.49}$$

When $\varphi(\cdot)$ is a linear function, $\delta_s = \delta_i = (1/2)$. For $\varphi(\cdot) \in C_m$, the class of (first and third quadrant) monotone functions, $0 < \delta_s \leq 1$ and for $\varphi(\cdot) \in C$, $0 < \delta_s \leq \infty$. Larger the value of δ_s (and smaller the value of δ_i), greater will be the arbitrariness of the nonlinear function $\varphi(\cdot)$. For using the stability criteria to be derived below for the nonlinear system (2.1), it appears advantageous to define sub-classes of nonlinear functions $\varphi(\cdot)$ in terms of prespecified values of δ_s and $(\delta_s - \delta_i)$.

We now present the first lemma on a nonnegative integral involving $\varphi(\cdot) \in C$ and a specific $Z(s)$. This sets the pattern for other lemmas involving $\varphi(\cdot) \in C_m$, C_{mo}.

Lemma 3.51. Let $m_o(s) = a_o + b_o s$, and $n(s) = 1$ for some constants $a_o > 0$, $b_o \geq 0$. Then the integral

$$I_o = \int_0^t \exp(\alpha_o \tau)\varphi(n_o(D)\sigma(\tau))(m_o(D)\sigma(\tau))d\tau + b_o \Phi(\sigma(0)) \tag{3.50}$$

is nonnegative for all $t \geq 0$ for $\varphi(\cdot) \in C$ if $\alpha_o \leq (a_o/b_o\delta_s)$.

Proof. We have

$$I_o = a_o \int_0^t \exp(\alpha_o \tau)\varphi(\sigma(\tau))\sigma(\tau)d\tau$$

$$+ b_o \int_0^t \exp(\alpha_o \tau)\varphi(\sigma(\tau))(d\sigma/d\tau)d\tau + b_o \Phi(\sigma(0)) \tag{3.51}$$

Integrate by parts the second term on the right hand side of (3.51) to get

$$I_o = \int_0^t \exp(\alpha_o \tau)(a_o - \alpha_o b_o \delta(\sigma))\varphi(\sigma)\sigma d\tau + b_o \exp(\alpha_o t) \Phi(\sigma(t)).$$

from which it is evident that I_o is nonnegative for all t in $[0,\infty)$ and σ if $\alpha_o \leq (a_o/b_o\delta_s)$. The lemma is proved.

Remark 3.51. Note that, in Lemma 3.51, if $\delta_s = \infty$, as it happens in general whenever $\varphi(\cdot) \in C$, then α_o has to be chosen as zero. When $\varphi(\cdot) \in C$, we need to use $k(t)I_o$ (in the place of $V_2(\chi, t)$ of Sec.3.4) along with $V_1(\chi, t)$ as a Lyapunov - Corduneanu function candidate for the system (2.1). It has not been possible to use any other class of multiplier functions, for instance, RL-RC functions, while establishing Lemma 3.51. Success in the use of more general multiplier functions has been achieved at the expense of restricting the class of nonlinearities $\varphi(\cdot)$ to C_m, C_{mo}. The following lemma gives one such version.

Lemma 3.52. Let $m_1(s) = s + \nu\mu$ and $n_1(s) = s + \mu$ for some constants $\mu > 0$, $0 \leq \nu < 1$. Further, let $\sigma(t) = (D+\mu)\sigma_1(t)$. Then the integral

$$I_1 = \int_0^t \exp(\alpha_1\tau)\ \varphi(n_1(D)\sigma(\tau))(m_1(D)\sigma(\tau))d\tau + \Phi(\sigma_1(0))\ /\ \mu(1-\nu) \qquad (3.52)$$

is nonnegative for all $t \geq 0$ for $\varphi(\cdot) \in C_m$ if $\alpha_1 \leq (\nu\mu/\delta_s)$.

Proof. The proof is similar to the proof of Lemma 3.51 and uses the property of monotone functions. See Exercise 3.10..

Corollary 3.521. Let $Z_2(s) = (m_2(s)/n_2(s)) = a_o + b_o s + \sum_i \gamma_i(s+\nu_i\mu_i)/(s+\mu_i)$,

where constants $a_o > 0$; $b_o, \gamma_i, \mu_i \geq 0$; $0 \leq \nu_i < 1$ for all i (finite) ; $\sigma_2(t) \triangleq n(D)\sigma(t)$; $(D + \mu_i)\sigma_3(t) \triangleq \sigma_2(t)$. Then the integral I_2 defined by

$$I_2 = \int_0^t \exp(\alpha\tau)\varphi(n_2(D)\sigma(\tau))\ (m_2(D)\sigma(\tau))d\tau + \Phi(\sigma_2(0)) + \sum_i \gamma_i\Phi(\sigma_3(0))/\mu_i(1-\nu_i)$$

is nonnegative for $\varphi(\cdot) \in C_m$ if $\alpha \leq \min_i(a_o/b_o, \nu_i\mu_i)/\delta_s$.

Proof. Combine the proofs of Lemmas 3.51 and 3.52.

Remark 3.52. The phase angle of the multiplier function $Z_1(s)$ in Lemma 3.52 lies in the band $[0, \pi/2]$, the maximal value being less than $\pi/2$ for nonzero ν. The same conclusion holds for Corollary 3.521 as well. The difference between the multiplier functions $Z_1(s)$ and $Z_2(s)$ is that phase angle of the latter can be made flat over a broader band of frequencies. Recalling that one of the purposes (while analyzing the stability of (2.1)) of employing a multiplier function $Z(s)$ is to satisfy the inequality $|\arg ZG(j\omega)| \leq (\pi/2)-\varepsilon$ for some constant $\varepsilon > 0$, we find that $Z_2(s)$ could be tried as a possible multiplier function candidate if the Nyquist plot of $G(j\omega)$ lies to the right of a line through the origin.

Remark 3.53. The inverse transform $z_1(t)$ of $Z_1(s)$ in Lemma 3.52 is given by

$$z_1(t) = 1 - (1-\nu)\mu\exp(-\mu t),\quad t \geq 0$$
$$= 0,\quad t < 0.$$

Hence the time varying part of $z_1(t)$ is negative for $t \geq 0$, and the integral of its modulus is less than 1. A similar interpretation is valid for $Z_2(s)$ of Corollary 3.521.

In Sec.3.4 dealing with linear system stability analysis, we used a general positive real function.Here we are confining ourselves to specific forms of multipliers. By considering biquadratic impedance functions Corollary 3.521 can be generalized. Still, we are short of using a general positive real function. We now state what is believed to be a general result motivated by Remark 3.53.

Lemma 3.53. Let $Z(s) \triangleq m(s)/n(s) = 1 + Z'(s)$ with $z'(t)$, the inverse transform of $Z'(s)$, identically zero for $t < 0$. Then with $\varphi(\cdot) \in C_m$ the integral

$$I_3 = \int_0^t \exp(\alpha\tau)\varphi(n(D)\sigma(\tau))(m(D)\sigma(\tau))d\tau$$

for arbitrary $\sigma(\cdot)$ for which the integral is defined, and for some nonnegative constant α, is nonnegative if

$$\text{(i)}\quad z'(t) \leq 0 \quad \text{for} \quad t \geq 0\ ;$$

$$\text{(ii)}\int_0^{\infty} \exp(\alpha t)|z'(t)|dt \leq 1/(1+\delta_s-\delta_i) \tag{3.53}$$

where δ_s and δ_i are defined by (3.49).

Proof . See Appendix 3.1.

Remark 3.54. Note that, in the statement of Lemma 3.53, no assumption as to the positive realness of $Z(s)$ was made. (But such an assumption will later be needed when actually establishing stability criteria for (2.1)). Further, the phase angle of $Z(j\omega)$ can lie in the band $[-\pi/2,\pi/2]$ with no limit to the number of switchovers from positive to negative values and vice versa. This contrasts significantly with the special multiplier of Corollary 3.521, and those in the literature.

Remark 3.56. Lemmas 3.52 and 3.53 (and Corollary 3.521) hold for $\varphi(\cdot) \in C_{mo}$ straightaway. In fact, weaker conditions on the multiplier functions are possible. For instance, in Lemma 3.53 hypothesis (i) requiring nonpositivity of $z'(t)$ is not needed. See Exercise 3.10 for integrals involving $\varphi(\cdot) \in C_{mo}$.

Based on the above preliminaries, exponential stability conditions for the system (2.1) with $\varphi(\cdot) \in C$ and C_m can be stated and proved. For the results concerning the case $\varphi(\cdot) \in C_{mo}$, see Exercises 3.10 and 3.12.

3.5(b) Derivation of Exponential Stability Criteria.

As before the state vector of (3.16) is denoted by $\underline{\chi}(t)$ of which the state vector $\underline{x}(t)$ of (2.1) is a subspace. Based on the positiverealness of $Z(s-\beta)G(s-\beta)$ for some constant $\beta > 0$, the positive definite quadratic form $V_1(\underline{\chi},t)$ is defined by (3.23). Now consider the integral

$$V_3(\underline{\chi},t) = \exp(-\alpha t)\int_0^t (\exp(\alpha\tau))\varphi(n(D)q(D)\chi(\tau))(m(D)q(D)\chi(\tau))d\tau \qquad (3.54)$$

for some constant $\alpha \geq 0$. Under suitable restrictions on $\varphi(\cdot)$ and $Z(s) \triangleq m(s)/n(s)$, $V_3(\underline{\chi},t)$ can be made nonnegative for all $\chi(\cdot)$ for which the integral (3.54) is defined.

Let

$$V_n(\underline{\chi},t) = V_1(\underline{\chi},t) + k(t)V_3(\underline{\chi},t) \qquad (3.55)$$

where $V_1(\underline{\chi},t)$ is defined by (3.23) and $V_3(\underline{\chi},t)$ by (3.54).

The time-derivative of $V_n(\underline{\chi},t)$ defined by (3.55) along the solutions of (3.16) is given by

$$\left.\frac{dV_n(\underline{\chi},t)}{dt}\right|_{(3.16)} = -2\beta V_1(\underline{\chi}(t),t)-(r_1(D+\beta)y(t))^2+((\tfrac{dk}{dt}/k)-\alpha)k(t)V_3(\underline{\chi}(t),t) \quad (3.56)$$

Where $\underline{\chi}(t)$ = solution vector of (3.16). From (3.56), we get the following differential inequality

$$(dV_n(\underline{\chi},t)/dt)\Big|_{(3.16)} \leq \sup_{t\geq 0}(-2\beta,\theta(t)-\alpha)V_n(\underline{\chi}(t),t) \qquad (3.57)$$

Observe that $V_n(\underline{\chi},t)$ satisfies inequality (3.22) of Corollary 3.331 with $\underline{x}$ replaced by $\underline{\chi}$. We can now state the conditions for exponential stability of (2.1) with $\varphi(\cdot)\in C_m$.

Theorem 3.51. The system (2.1) is exponentially stable if (i) there exists a positive real function $Z(s) \triangleq a_o + b_o s + Z_c(s)$ where constants $a_o > 0$, $b_o \geq 0$, $Z_c(s) \triangleq m_c(s)/n_c(s)$ is not necessarily positive real but has an inverse transform $z_c(t)$ vanishing identically for $t < 0$ and nonpositive for $t \geq 0$, such that (ii) $Z(s-\beta)G(s-\beta)$ is positive real for some $\beta > 0$; $m(s-\beta)q(s-\beta)$(or $n(s-\beta)p(s-\beta)$)

and $(\text{Ev } q(s-\beta)m(s-\beta)p(-s-\beta)n(-s-\beta))^{(-)}$ have no common factors on the imaginary axis. (iii) for some constant $\alpha > 0$

$$\int_0^{\infty} \exp(\alpha t) \left| z_c(t) \right| dt \leq a_o'/(1 + \delta_s - \delta_i) \tag{3.58}$$

for some positive constant $a_o' < a_o$; (iv) for $\alpha' = \max \alpha$ of hypotheses (iii) and

$$\gamma = \min\left((a_o - a')/b_o\delta_s,\ \alpha'\right), \tag{3.59}$$

$$(1/T) \int_{t_o}^{t_o+T} (\theta(\tau) + 2\beta-\gamma)^+ d\tau \leq 2\beta - \nu \tag{3.60}$$

for all $T > 0$ and for some constant $\nu > 0$.

Proof. As a Lyapunov-Corduneanu function candidate for (3.16) (and hence for (2.1)) choose $V_n(\underline{\chi}, t)$ as defined by (3.55) with $V_1(\underline{\chi}, t)$ and $V_3(\underline{\chi}, t)$ given by (3.23) and (3.53) respectively, where $n(s) = n_c(s)$ and $m(s) = (a_o + b_o s)n_c(s) + m_c(s)$. Evidently,

$$V_3(\underline{\chi}, t) = \exp(-\alpha t)\int_0^t \exp(\alpha\tau)\varphi(n_c(D)q(D)\chi(\tau))((a_o+b_oD)n_c(D)+m_c(D))q(D)\chi(\tau)d\tau$$

$$= \exp(-\alpha t)\int_0^t \exp(\alpha\tau)\varphi(n_c(D)q(D)\chi(\tau))(a_o-a'+b_oD)n_c(D)q(D)\chi(\tau)d\tau$$

$$+ \exp(-\alpha t)\int_0^t \exp(\alpha\tau)\varphi(n_c(D)q(D)\chi(\tau))\,(m_c(D)q(D)\chi(\tau)+a'n_c(D)q(D)\chi(\tau))\,d\tau \tag{3.61}$$

for some positive constant $a' < a_o$.

From Lemma 3.51, the first integral on the right hand side of (3.61) is nonnegative if $\alpha \leq (a_o - a')/b_o\delta_s$; the second integral on the right hand side of (3.61) is nonnegative if hypotheses (i) and (ii) of Lemma 3.53 are satisfied. Consequently, $V_3(\underline{\chi}, t)$ is nonnegative if α were chosen as the minimum of $(a_o - a')/b_o\delta_s$ and α of hypothesis (iii). The time-derivative of $V_n(\underline{\chi}, t)$ along the solutions of (3.16) satisfies the inequality (3.57) where α is to be replaced by γ defined by (3.59).

Invoking Corollaries 3.331 and 3.332, we conclude that hypothesis (iv) guarantees exponential stability of (2.1).

Remark 3.57. Hypothesis (iv) of Theorem 3.51 can be improved, as in the case of the linear system (Remark 3.41), to the following form :

$$\text{(v)} \qquad (1/T)\int_{t_o}^{t_o+T} (\theta(\tau) + 2\beta-\gamma)^+ d\tau \leq M < \infty$$

for all finite $T > 0$ and for some constant $M>0$; (3.62)

$$\lim_{T\to\infty} (1/T)\int_{t_o}^{t_o+T} (\theta(\tau) + 2\beta-\gamma)^+ d\tau \leq 2\beta-\nu$$

for some constant $\nu > 0$.

An important special case of Theorem 3.51 is the following theorem.

Theorem 3.52. The system (2.1) with $\varphi(\cdot)\in C$ is exponentially stable if there exists a multiplier function $Z(s) = a_o + b_o s$ with constants $a_o > 0$ and $b_o \geq 0$, such that

(i) $\mathrm{Re}(a_o + b_o(j\omega-\beta))G(j\omega-\beta) \geq 0$ for all real ω and for some $\beta > 0$, and

(ii) for some nonnegative constant $\gamma \leq (a_o/b_o\delta_s)$, the set of integral inequalities (3.62) is satisfied.

Remark 3.58. If hypothesis (i) of Theorem 3.52 is satisfied for $b_o = 0$, then hypothesis (ii) can be dispensed with, implying thereby that no restriction need be placed on $\theta(t)$. The theorem is then equivalent to the circle criterion.(Use the results of Exercises 3.4 and 3.15)..

3.5 c. Geometric interpretation.

Case 1. $\varphi(\cdot)\in C$ for which $0 < \delta_s \leq \infty$.

Hypothesis (i) of Theorem 3.52 is identical with the Popov hypothesis [P 8(a)] for the time invariant nonlinear system (2.24) except that a 'shifted' $G(j\omega)$ is used .. Let $\mathrm{Re}\, G(j\omega-\beta) = X$ and $\mathrm{Im}\, G(j\omega-\beta) = Y$. Then hypothesis (i) of Theorem 3.52 implies the inequality, $\omega Y \leq ((a_o/b_o)-\beta)X$ for all real ω . If this inequality is to be satisfied, it is necessary and sufficient that the modified frequency characteristic of $G(j\omega-\beta)$ should lie to the right of a straight line through the origin. If this line has slope β_1, then $(a_o/b_o) = \beta + \beta_1$. Hence we have the following result.

Criterion 3.51. The system (2.1) is exponentially stable for $\varphi(\cdot)\in C$ if

(i) the modified frequency characteristic of $G(j\omega-\beta)$ lies to the right of straight line through the origin for some constant $\beta > 0$; (ii) with $\beta^* = \max \beta$ of hypothesis (i), and β_1 = the slope of the line so drawn , the set of integral inequalities (3.62) is satisfied for $\gamma = (\beta^* + \beta_1)/\delta_s$.

If the problem involves a finite nonlinear time varying gain, then a straight line through a point on the negative real axis is to be drawn. Refer to Exercises 3.15 and 3.25.

Case 2. $\varphi(\cdot) \in C_m$ for which $0 < \delta_s \leq 1$.

Observe that the argument of the multiplier function of Corollary 3.521 always lies in the band $[0, \pi/2]$, and the multiplier function could be constructed as to have a flat phase-frequency characteristic of the desired magnitude (F) within that band. The reciprocal of the multiplier of Corollary 3.521 could also be used to guarantee the nonnegativity of the I_2 (Exercise 3.12).In this case, the phase-frequency characteristic of the multiplier function could be made to be flat in the band $[-\pi/2, 0]$ of a desired magnitude (F). Recall that a multiplier function Z(s) is to be so chosen that the integral of the type I_2 (Corollary 3.521) involving the nonlinearity is nonnegative, and that $-\pi/2 + \varepsilon \leq \arg \{G(j\omega-\beta)Z(j\omega-\beta)\} \leq \pi/2 - \varepsilon$ for all real ω , for some positive constants β and ε. We conclude that,if $\arg\{G(j\omega-\beta)\}$ lies to the right of a straight line through the origin inclined at an angle F to the vertical axis, a multiplier of the type $Z_2(s)$ (in Corollary 3.521) or its reciprocal can be employed.

If $Z(s-\beta)G(s-\beta)$ is positive real for some $\beta > 0$, then $Z(s-\alpha)$ is positive real for some $\alpha \geq \beta$. Note that in Corollary 3.521 α could be chosen as $\min_i(a_o/b_o, \nu_i\mu_i)/\delta_s$ for nonnegativity of I_2. If the Nyquist plot of $G(j\omega-\beta)$ for some $\beta > 0$ lies to the right of a straight line through the origin, then $\min_i(a_o/b_o, \nu_i\mu_i)$ is at least equal to β. Therefore, we have the following geometric version of Theorem 3.51.

Criterion 3.52. The system (2.1) with $\varphi(\cdot) \in C_m$ is exponentially stable if (i) The Nyquist plot of $G(j\omega-\beta)$ for some $\beta > 0$ lies to the right of a straight line through the origin ; (ii) the set of integral inequalities (3.62) is satisfied for $\gamma = (\beta/\delta_s)$ where β is the maximal value permitted in hypothesis (i).

Remark 3.59. For a finite gain problem, Criterion 3.52 can be transformed to yield a result of the type of the circle criterion, but with the centre of the circle not on the negative real axis. See Exercises 3.4 and 3.15.

Remark 3.60. Theorem 3.51 has been inspired by the contribution of O'Shea [O 2(b)] for the time- invariant nonlinear system (2.24). However, the multiplier function Z(s) of the theorem is still required to be positive real for generating the positive definite quadratic form $V_1(\underline{\chi},t)$ defined by (3.23). The part $Z_c(s)$ of Z(s) need not be positive real. In order to derive a geometric version of Theorem 3.51, we have to construct first a positive real (rational) multiplier function from the phase angle-frequency characteristic of $G(j\omega-\beta)$ and then express the multiplier function in the form given in the theorem statement. Verification of hypothesis (iii) involving the time domain bound on the multiplier function is the more difficult part. An open problem is to express hypothesis (iii) in terms of the phase angle-frequency characteristic of $G(j\omega-\beta)$.

3.6. Comparison with Other Results.

(i) The first stability result on nonlinear time varying system is due primarily to Zames [3(a)], Sandberg [S 2(a)] and Narendra and Goldwyn [N 1]. This result is the well known circle criterion. As indicated above, Theorem 3.52 for $b_o = 0$ is equivalent to the circle criterion (Exercise 3.4). In this case, Theorem 3.52 involves merely the frequency domain inequality $\text{Re}\, G(j\omega-\beta) \geq 0$ for all real ω, for some $\beta > 0$. Note that this condition is severe as it amounts to the requirement of positiverealness of $G(s-\beta)$. Further, for the linear time invariant system (2.15), Theorem 3.52 for $b_o = 0$ does not reduce to the Nyquist criterion.

The inference is that a transition from stability results of the type of Theorem 3.52 to the Nyquist criterion is plausible by restricting the rate of variation of the time varying gain. With this, naturally, the requirements on $G(j\omega-\beta)$ can be weakened. A result on these lines is due to Zames [Z 3(c)] where a constraint of the type $\theta(t) \lesssim$ a constant is derived for $\varphi(\cdot) \in C_m$ on choosing an RC-multiplier function. Other results, for instance, those found in [N 2] employ special classes of positive real multipliers in the attempt to weaken the requirements on $G(j\omega)$.

ii) Lemma 3.33 (due to Corduneanu) has also been used in [N 2] for deriving asymptotic stability (Definition 2.32) conditions which are difficult to verify and difficult to compare with others in literature. The method of generating suitable Lyapunov function candidates involves multiple integrals of the type

$$\sum_i \beta_i \int_0^{p_i(D)y} \varphi(r_i(D)y)\,d(r_i(D)y)$$

where $p_i(s)$ and $r_i(s)$ are obtained from $Z(s)$ and $G(s)$. For conversion of the matrix inequalities (in the terminology of Yakubovich [Y 1(a)]) so obtained into frequency domain form, we must have recourse to the classic Kalman-Yakubovich-Popov lemma [K 1(b)] [Y 1(a)] [P 8(a)] . It is not known, for instance, how to derive stability conditions of the type of the Theorem 3.51, using the method of [N 2].

iii) The stability criteria of Haddad [H 2] derived from certain transformations and time domain inequalities do not give anything new (Exercise 3.8). It is not known whether the classical stability boundary for the Mathieu equation can be reproduced.

iv) The construction of a multiplier function given by Freedman [F 5(a)] for the time invariant nonlinear system (2.24) is not wholly correct for the same reason as explained in Sec 3.4(b).

v) Extension of the 'spectral radius' concept of Zames and Kallman [Z 5] to nonlinear periodic coefficient systems as found in [V 3] does not give, when applied to Mathieu equation, the well known stability boundary. The results of the present chapter also cannot incorporate explicitly the information of periodicity of $k(t)$ and hence are definitely not the best of their type. Common with all the criteria in the literature, for systems of an order higher than 2, these results make sense only for 'damped' systems, i.e., for $\beta > 0$. Considerable work remains to be done concerning stability of (linear and) nonlinear systems with no inherent damping. This will hopefully pave the way for an improvement in the results of the chapter.

3.7. Conclusions.

What emerges from an application of the Lyapunov-Corduneanu theorem (Lemma 3.33 and its corollaries) to the exponential stability analysis of the nonlinear time varying feedback system (2.1) is that a damped linear time invariant system (2.15) retains, for time varying feedback gain, its exponential stability property if (i) in the case of the linear system (2.21), the normalized rate of variation $\theta(t)$ of $k(t)$ satisfies a certain global bound depending on the amount of damping β in the time invariant system (2.15) ; and (ii) in the case of the nonlinear system (2,1), $\theta(t)$ satisfies a similar global bound dependent on β and the class of $\varphi(\cdot)$. Still the basic stability problems remain unsolved. This is in view of the inadequacy of the results of the chapter.

(i) For the (undamped) Mathieu equation, Theorem 3.41 or Theorem 3.42 gives a trivial result.

(ii) A geometric interpretation of Theorem 3.51 for the nonlinear time varying system (2.1) with $\varphi(\cdot) \in C_m$ is not known.

(iii) Suppose $k(t)$ is periodic with period T_0. It would be useful to derive criteria of the type of Theorems 3.41 and 3.51 with T_0 incorporated in the conditions. Note the classical stability criteria [C 2] for linear periodic systems even without damping do not involve the rate of variation of $k(t)$.

(iv) In the absence of exponential stability 'in the large' (see Remark 2.33 in Chapter 2) or of damping, possibly a finite region in the state space may be chosen and exponential stability of the system established in this region. (Exercise 3.26).

The methods of Popov and Zames-Sandberg in the next two chapters lead to some improvements in the stability conditions. Specifically, (i) for a periodic $k(t)$, with period T_0, stability conditions are expressed in terms of T_0 ; and (ii) the multiplier function can be a combination of causal and anticausal functions resulting in flexible lower and upper global bound on $\theta(t)$. However, the last word on the stability problem has not yet been said.

Appendix 3.1

Proof of Lemma 3.53. Let $n(D)\sigma(\tau) \triangleq \sigma_1(\tau)$. We have

$$I_3 = \int_0^t \exp(\alpha\tau)\varphi(\sigma_1(\tau))\sigma_1(\tau)d\tau + \int_0^t \exp(\alpha\tau)\varphi(\sigma_1(\tau)) \left(\int_0^\infty z(\tau')\sigma_1(\tau-\tau')d\tau'\right)d\tau \quad (3.63)$$

Assuming that the order of integration in the last integral of (3.63) can be interchanged, we get, after some minor manipulations,

$$I_3 = \int_0^t \exp(\alpha\tau)\varphi(\sigma_1(\tau))\sigma_1(\tau)d\tau + \int_0^\infty z'(\tau')\exp(\alpha\tau') \left(\int_0^t \varphi(\sigma_1(\tau))\sigma_1(\tau-\tau')\exp(\alpha(\tau-\tau'))d\tau\right)d\tau' \quad (3.64)$$

From the monotone property of $\varphi(\cdot)$ we have

$$\varphi(y_1)(y_1-y_2) \geq \Phi(y_1) - \Phi(y_2) \text{ for all } y_1 \text{ and } y_2, \quad (3.65)$$

where $\Phi(y_1) = \int_0^{y_1} \varphi(\sigma_1)d\sigma_1$.

Define $y_1 = \sigma_1(\tau)$ and $y_2 = \sigma_1(\tau-\tau')\exp(-\alpha\tau')$. (3.66)

Using (3.65), we can write the following inequality

$$\int_0^t \exp(\alpha\tau)\varphi(\sigma_1(\tau))(\sigma_1(\tau)-\sigma_1(\tau-\tau')\exp(-\alpha\tau'))d\tau \geq$$

$$\int_0^t \exp(\alpha\tau)\,\Phi(\sigma_1(\tau))d\tau - \int_0^t \exp(\alpha\tau)\,\Phi(\sigma_1(\tau-\tau')\exp(-\alpha\tau'))d\tau. \quad (3.67)$$

The last integral of (3.67) can be rewritten by changing the variable of integration to $\tau_1 = \tau-\tau'$:

$$\int_0^t \exp(\alpha\tau)\Phi(\sigma_1(\tau-\tau')\exp(-\alpha\tau'))d\tau = \int_{-\tau'}^{t-\tau'} \exp(\alpha(\tau_1+\tau'))\Phi(\sigma_1(\tau_1)\exp(-\alpha\tau'))d\tau_1$$

But $\Phi(\sigma_1(\tau_1)\exp(-\alpha\tau')) \leq \delta_s\varphi(\sigma_1\exp(-\alpha\tau'))\sigma_1(\tau_1)\exp(-\alpha\tau')$ and $\Phi(\sigma_1(\tau)) \geq \delta_i\varphi(\sigma_1(\tau))\sigma_1(\tau)$ from (3.49). Therefore, noting that $\sigma_1(\tau_1) = 0$ for $\tau_1 < 0$, the right hand side of inequality (3.67) is greater than or equal to

$$\delta_i \int_0^t \exp(\alpha\tau)\varphi(\sigma_1(\tau))\sigma_1(\tau)d\tau - \delta_s \int_0^t \exp(\alpha\tau)\varphi(\sigma_1(\tau)\exp(-\alpha\tau'))\sigma_1(\tau)d\tau \quad (3.68)$$

Once again, from the monotonicity of $\varphi(\cdot)$ we have

$$\varphi(\sigma_1(\tau)\exp(-\alpha\tau'))\sigma_1(\tau) \leq \varphi(\sigma_1(\tau))\sigma_1(\tau)$$

which, when used in (3.68) and subsequently in (3.67), gives

$$\int_0^t \exp(\alpha\tau)\varphi(\sigma_1(\tau))(\sigma_1(\tau-\tau')\exp(-\alpha\tau'))d\tau \leq (1+\delta_s-\delta_i) \int_0^t \exp(\alpha\tau)\varphi(\sigma_1(\tau))\sigma_1(\tau)d\tau.$$

Consequently, from (3.64) we conclude that I_3 is nonnegative if $z'(\tau') \leq 0$ for all $\tau' \geq 0$ and the time domain inequality (3.53) is satisfied.

If $\varphi(\cdot) \in C_{mo}$, i.e., $\varphi(\cdot) \in C_m$ and $\varphi(\sigma) = -\varphi(-\sigma)$ for all $\sigma \neq 0$, then write the monotone inequality (3.65) for $-y_2$ and carry through the above calculation to infer

$$\left| \int_0^t \exp(\alpha\tau)\varphi(\sigma_1(\tau))(\sigma_1(\tau-\tau')\exp(-\alpha\tau'))d\tau \right| \leq (1+\delta_s-\delta_i) \int_0^t \exp(\alpha\tau)\varphi(\sigma_1(\tau))\sigma_1(\tau)d\tau,$$

which in association with (3.64) gives the result that Lemma 3.53 is valid for $\varphi(\cdot) \in C_{mo}$ without hypothesis (i).

EXERCISES

3.1. Find the Fourier transform (or bilateral Laplace transform) of the time functions defined by

(i) $z(t) = \sum_i \alpha_i \exp(-\gamma_i t) + \delta_D(t)$ for $t \geq 0$

$= 0$ for $t < 0$;

(ii) $z(t) = \sum_i \alpha_i \exp(\gamma_i t) + \delta_D(t)$ for $t \leq 0$

$= 0$ for $t > 0$;

(iii) $z(t) = \sum_i \alpha_i \exp(-\gamma_i t) + \delta_D(t)$ for $t \geq 0$,

$$= \sum_i \alpha_i' \exp(\gamma_i' t) \quad \text{for} \quad t < 0 ;$$

$$\text{(iv)} \; z(t) = \sum_i \alpha_i \exp(-\gamma_i t) \cos(\eta_i t + \mu_i) + \delta_D(t) \; \text{for} \; t \geq 0,$$

$$= \sum_i \alpha_i' \exp(\gamma_i' t)\cos(\eta_i' t + \mu_i') \; \text{for} \; t < 0,$$

where γ_i and $\gamma_i' > 0$ for all i, and $\delta_D(t)$ is the Dirac delta function. Identify causal, anticausal and noncausal functions. Plot their phase angle vs frequency characteristics. What are the conditions under which these functions could be used to define path integrals independent of the path ?

3.2. Given the bilateral Laplace transform of a time function as

$$Z(s) = 1 + \sum_i \alpha_i/(s+\gamma_i) + \sum_{i'} \beta_{i'}/(-s+\mu_{i'})$$

According to Willems [W 3(b), p 647], this could be interpreted as the transform of an impulse response which does not vanish for negative time or as the transform of an impulse response whose integral diverges. Use both the interpretations and check whether, under the real part condition Re $Z(j\omega) \geq 0$ for all real ω, path independent integrals could be defined leading to positive definite, negative definite or indefinite quadratic forms.

3.3(a). Let $z_{ac}(t)$ be defined nonzero for $t \leq 0$ and identically zero for $t > 0$. Let $f(\cdot)$ belong to the class of functions for which the following integral makes sense.

$$I_{ac} = \int_0^t f(\tau)\left(\int_{-\infty}^0 z_{ac}(\tau')f(\tau-\tau')d\tau'\right)d\tau$$

What are the conditions for I_{ac} to be nonnegative for all $t \geq 0$ for arbitrary $f(\cdot)$ in the chosen class of functions ? Suppose $z_{ac}(t)$ has a rational transform, satisfying the inequality Re $Z_{ac}(j\omega) \geq 0$ for all real ω, can one generate path independent integrals which have a sign definite property ?

3.3(b). Suppose it is given that Re $Z(j\omega) \geq 0$ and Re $Z(j\omega)G(j\omega) \geq 0$ for all real ω. Poles of $G(s)$ are in the open left half of the complex plane. No restrictions are imposed on the location of the poles of $Z(s)$. Can one generate positive definite quadratic functions by means of path independent integrals ?

3.4. Suppose that $\mathrm{Re}(Z(j\omega)+(1/K_1))/(Z(j\omega)+(1/K_2)) \geq 0$ for all real ω for some constants K_1 and K_2 with $K_1 < K_2$. Then show that the Nyquist locus of $Z(j\omega)$ does not intersect the open disk in the G-plane centred at the point $-(K_1+K_2)/2K_1K_2$ and having a radius $(K_2-K_1)/2K_1K_2$.

3.5. Show that the LC multiplier generated by Brockett and Willems [B 20, Theorem 2, p 257] can be obtained as special case of the construction outlined in Lemma 3.22.

3.6. As $k(\cdot) \in \mathcal{K}$, the class of functions given by Definition 3.41, the non-linear time varying system (2.1) can as well be described by

$$(1/k(t))p(D)y+\varphi(q(D)y) = 0 \quad \text{for } t \geq 0.$$

Choosing a Lyapunov-Corduneanu function candidate of the form $(1/k(t))V_1(\underline{\chi},t)+(V_2(\underline{\chi},t)$ or $V_3(\underline{\chi},t))$ where $V_1(\underline{\chi},t)$, $V_2(\underline{\chi},t)$ and $V_3(\underline{\chi},t)$ are defined respectively by (3.23), (3.24) and (3.54), derive exponential stability conditions for the linear and nonlinear time varying systems (2.21) and (2.1).

3.7. Find the minimum and maximum eigenvalues of the system matrix for the differential equation

$$\begin{bmatrix} \dfrac{dx_1}{dt} \\ \\ \dfrac{dx_2}{dt} \end{bmatrix} = \begin{bmatrix} 0 & 1 \\ \\ -\eta-\gamma\cos 2t & -\alpha \end{bmatrix} \begin{bmatrix} x_1 \\ \\ x_2 \end{bmatrix}$$

for positive constants η, γ and α. Use $(x_1^2 + x_2^2)$ as a Lyapunov-Corduneanu function candidate and derive exponential stability criteria using the maximum and minimum eigenvalues. Compare with the result obtainable from Theorem 3.42.

3.8. Apply Theorems 3.41 and 3.42 to the lightly damped Mathieu equation

$$\frac{d^2y}{dt^2} + 2\mu\frac{dy}{dt} + (\mu^2+a)y + 16q(\cos 2t)y = 0, \quad t \geq 0$$

where $0 < \mu << a$. Compare with the classical stability boundary as found in McLachlan [M 7] and Hayashi [H 6], and also with that of Zames and Kallman [Z 5]

Analyse the case of $\mu = 0$. Apply the criteria of Willems [W 3(a)], Haddad [H 2], Davis [D 3(a)] and Vidyasagar [V 3], and comment on their effectiveness.

3.9. Find the forms or give examples of nonlinearities $\varphi(\cdot) \in C$, C_m with specified δ_s and $\delta_s - \delta_i$.

3.10(a). Prove Lemma 3.52.

3.10(b). Let $Z(s) \triangleq m(s)/n(s) = a_o + b_o s + \sum_i \gamma_i (s + \nu_i \mu_i)/(s+\mu_i)$ where $a_o > 0$; b_o, $\gamma_i, \mu_i \geq 0$; and $1 \leq \nu_i < 2$ for all i. Find the conditions for the integral

$$I = \int_0^t \exp(\alpha\tau)\varphi(n(D)\sigma(\tau))(m(D)\sigma(\tau))d\tau$$

with $\varphi(\cdot) \in C_{mo}$ to be nonnegative (except for terms of the type found in Corollary 3.521 to Lemma 3.52).

3.11. Suppose $Z(s) = m(s)/n(s)$ as defined in Ex.3.10. Find conditions for the integral

$$I = \int_0^t \exp(\alpha\tau)\varphi(m(D)\sigma(\tau))(n(D)\sigma(\tau))d\tau$$

with $\varphi(\cdot) \in C_m$ to be nonnegative (except for terms of the type found in Corollary 3.521 to Lemma 3.52).

3.12. For $Z(s)$ as defined in Corollary 3.521 to Lemma 3.52, repeat Ex.3.11 with $\varphi(\cdot) \in C_{mo}$.

Based on Ex.3.11 and the present result, can one conclude that the reciprocal of the multiplier of Ex.3.10 can be used to establish Corollary 3.521 to Lemma 3.52 and vice versa.

3.13. Define a new class of nonlinearities - the class of power law nonlinearities - as follows [B 20], [T 3] : $\varphi(\cdot) \in C_{mo}$ and

$$\left|\sigma_1/\sigma_2\right|^{1/m_o} \leq \left|\varphi(\sigma_1)/\varphi(\sigma_2)\right| \leq \left|\sigma_1/\sigma_2\right|^{m_o} \quad \text{for } |\sigma_1| \geq |\sigma_2|$$

and $m_o \geq 1$, or equivalently

$$\varphi(\sigma_1)(c\sigma_1+\sigma_2)+\varphi(\sigma_2)(c\sigma_2-\sigma_1) \geq 0 \quad \text{for all } \sigma_1 \text{ and } \sigma_2 \text{ where}$$

$$c = \max_{0<\sigma<\infty} \left| \frac{\sigma^{m_o} - \sigma}{1+\sigma^{m_o+1}} \right| .$$

Note that as $m_o \to \infty$, $c \to 1$, so that $\varphi(\cdot) \in C_{mo}$; and as $m_o \to 1$, $c \to 0$ giving a linear function ; $c = 0.3536$ corresponds to $m_o = 3$. Derive results corresponding to Lemmas 3.52 and 3.53, and Theorem 3.51. Can we get Theorem 3.41 as a special case for $c = 0$, and the Nyquist criterion when $k(t) = K$, a positive constant ?

3.14. Check whether Lemma 3.53 includes Corollary 3.521 (Lemma 3.52) as a special case. If not, how would you improve Lemma 3.53 ? Note that, for $\alpha = 0$, Lemma 3.53 does not reduce to the result of O'Shea [O 2(b)] derived in the context of time invariant nonlinear systems ? Suggest means of rectifying the situation.

3.15. Use the results of Ex.2.2 to derive the finite gain version of Theorems 3.41, 3.42, 3.51 and 3.52, and of Criteria 3.51 and 3.52 in Sec.3.5(c).

3.16. Apply Theorem 3.41 or 3.42 to determine exponential stability conditions for the system

$$\frac{d^2y}{dt^2} + 3\frac{dy}{dt} + (2-\gamma \sin t)y = 0, \quad t \geq 0.$$

Compare with the bound $|\gamma| < 8.74$ of Gunderson et al [G 7] . Write the system in matrix form and using the well known formula

$$\text{determinant of the system transition matrix} = \exp\left(\int_{t_o}^{t} (\text{Trace of system matrix})d\tau\right),$$

comment on stability (or instability) conditions so obtainable.

3.17. For the linear time varying system (2.21) with $G(s) = s/(s+10)(s^2+0.4s+1)$ derive conditions for exponential stability for (i) a periodic $k(t)$ with finite gain and (ii) nonperiodic $k(t)$ with large gain. Compare with the results of Willems [W 3(a)].

3.18. For the linear time varying system (2.21) with $G(s) = s^2/(s^3+2s^2+2s+1)$, [B 17] check whether Theorems 3.41 and 3.42 (and the finite gain versions obtained in Ex.3.15) can be applied. If the answer is no, suggest means of overcoming the drawback.

3.19. For the nonlinear time varying system with $G(s) = s^2/(s^4+3s^3+4s^2+2s+1)$ under a suitable choice of the class of nonlinearities, derive conditions for exponential stability. Use the results of Ex.3.15.

3.20. Find conditions for the exponential stability of (2.1) with $G(s) = (s+10)^3/(s+1)^2(s+100)$ and $\varphi(\sigma) = (1/2)\tan^{-1}\sigma$ [Z 3(c)].

3.21. The system (2.1) with $G(s) = (-s+1)/s^2+3)$ is unstable if $k(t)\varphi(\cdot)$ were replaced by a constant $K \geq 0$. Naturally, the theorems of the chapter cannot be used to derive exponential stability conditions for (2.1). Suggest new methods for tackling the stability problem of such a system.

3.22. The system (2.1) with $G(s) = (1/(s^3+s^2+s))$, and $k(t) = K(1+\exp(-t))$; $\varphi(\sigma) = \sigma(9\sigma^2+10)/10(\sigma^2+1)$ is to be analysed for exponential stability. Find the maximum value of K for exponential stability and compare with the Hurwitz sector value for the time invariant system (2.15).

3.23. For the system (2.1) with $G(s) = (\alpha s^2+1)^2/(\gamma_2 s^2+\gamma_1 s+1)^3$ where α, γ_2 and γ_1 are positive constants, determine the Hurwitz sector and check for the applicability of Theorems 3.51 and 3.52. This is the system, absolutely stable for time invariant feedback and suitable choice of constants α, γ_1 and γ_2 for which according to Pyatnitskii [P 10(b)], the Popov criterion fails.

3.24. For the system (2.1) with $G(s) = (10s+1)(2s+1)/(s^2+20s+400)(2s^2+5s+4)$ [B 20], plot the phase angle-frequency characteristic of G(s). Based on Lemma 3.22 and on the construction of a positive real function from its phase angle as found, for instance, in Guillemin [G 6, p 316], determine conditions for exponential stability.

3.25. Consider the system (2.1) with $G(s) = (s+2)/(s+1)(s+1.5)(s^2+4s+12)$ and $k(t), \varphi(\cdot)$ satisfying the inequalities

$$0 < \varepsilon \leq k(t) \leq 1 \quad \text{for all} \quad t \geq 0 ;$$

$$0 < \varepsilon \leq (\varphi(\sigma)/\sigma) \leq K_N \quad \text{for all} \quad \sigma \neq 0,$$

for some unknown constant K_N. With $\varphi(\cdot) \in C$ and C_m, determine the maximum values of K_N for exponential stability.

3.26. The Lyapunov-Corduneanu method provides the means of investigating a system's exponential stability properties in a subset of the state space. Consider the system

$$\frac{d^2y}{dt^2} + \frac{dy}{dt} + y + k(t)y^2 = 0, \quad t \geq 0$$

where $k(t)$ is nonnegative and uniformly bounded on $[t_o, \infty)$. Find the region (in the state space) of exponential stability for the system.

3.27. Suppose for the system (2.1), $G(s) = (s-1)/(s+1)^3$ and $k(t)\varphi(\sigma) = K\,\sigma(1+\cos\sigma)\sin t/(1+100\,|\sin t|)$ with $K > 0$.

Find the largest value of K for which the system is exponentially stable. Compare the result with $K = (1/2)$ given by the circle criterion and $K = 10$ given by Haddad [H 2].

3.28. In 1836 Sturm (see [E 2] and [C 6]) posed the problem : For which $k_1(t)$ and $k_o(t)$ will the solutions $y(t) \neq 0$ (in the class of twice differentiable functions) of the equation

$$\frac{d^2y}{dt^2} + k_1(t)\frac{dy}{dt} + k_o(t)y = 0, \quad t \geq 0 \tag{3.68}$$

where $k_1(t)$ and $k_o(t)$ are continuous on $t \geq t_o \geq 0$ be oscillatory, i.e., will each of them have an infinite succession of zero values ?

According to El'sin [E 2], the solutions of (3.68) will be oscillatory if and only if

$$-(k_1(t)/2) + \int_{t_o}^{t} (k_o(\tau)-(k_1^2(\tau)/4))d\tau = k_2(t)-k_3(t)$$

where $k_2(t)$ is differentiable and $k_3(t)$ is continuous, such that

$$\int_{t_o}^{\infty} \exp\left(-2\int_{t_o}^{\tau'} k_3(\tau)d\tau\right)d\tau' = \infty$$

$$\int_{t_o}^{\infty} \exp\left(2\int_{t_o}^{\tau} k_3(\tau')d\tau'\right)\left(\frac{dk_2(\tau)}{d\tau} + k_3^2(\tau)\right)d\tau = \infty$$

Let $k_1(t) = 0.01$ and $k_o(t) = a+b\cos 2\omega t$. Pick a and b on the boundary of the stability region given by Hayashi [H 6] Zames and Kallman [Z 5]. Can one check the existence of oscillations for these values ? Apply the method to the case of $k_1(t) \equiv 2$ and arbitrary $k_o(t)$ and compare with the results of Brockett [B 16(c)(d)].

3.29. A graphical interpretation of a particular case of Ex.3.28 is as follows. Consider the system

$$\frac{d^2 y}{dt^2} + a\frac{dy}{dt} + k(t)y = 0, \quad t \geq 0. \tag{3.69}$$

Assume that constant $a > 0$ and $k(t) \geq 0$ for all $t \geq 0$. On the phase plane (dy/dt) vs y, locate the point $(0,\gamma_1)$ on the (dy/dt)-axis where $\gamma_1 > 0$. Find the time function $k(t)$ belonging to the class of absolutely continuous functions such that (i) the trajectory cuts the y-axis in finite time T_1 such that the intercept γ_2 is maximal ; and (ii) the subsequent trajectory cuts the (dy/dt)-axis at $(0, -\gamma_1)$ in finite time T_2. Naturally, by symmetry, the strategy of choosing $k(t)$ in the first quadrant can be applied in the third quadrant to result in maximal intercept γ_2 on the negative y-axis. Time taken for the trajectory to reach $(-\gamma_2,0)$ from $(0,-\gamma_1)$ is T_1. Finally the original point $(0,\gamma_1)$ is reached in finite time T_2 by a choice of $k(t)$ adopted in the fourth quadrant. The total time taken for the trajectory to close on to itself in such a way that γ_2 is maximal, is finite and equal to $2(T_1+T_2)$. The above choice of $k(t)$ is optimal in the sense that a different $k(t)$ would either result in an unbounded trajectory or a trajectory which tends to the origin. It is quite likely that there are many forms of $k(t)$ which give rise to a closed loop trajectory. Let this class of functions $k(t)$ be denoted by $\mathcal{K}_o$. An increase in the bound on $k(t) \in \mathcal{K}_o$ will result in an unbounded trajectory and a decrease in the bound on $k(t)$ in an asymptotically stable trajectory. A solution to the problem does exist : The lightly damped Mathieu equation where $a = 0.01$ and $k(t) = \alpha + \beta \cos\omega t$, α and β being chosen to lie on the stability boundary given in McLachlan [M 7] or Hayashi [H 6]. Note that the solution given by Brockett [B 16(c)(d)] and repeated in Willems [W 4(c),p 115-117] is faulty because the strategy of choosing $k(t)$ as constant only enables the trajectory, starting from $(0,\gamma_1)$, to cut the y-axis at $(\gamma_2,0)$ in infinite time. Compare with the results given for $a = 0$ in (3.69) by Lyapunov, Borg and Krein [S9] who do not even assume that $k(t) \geq 0$ for all $t \geq 0$. See also Bermant and Yemelyanov [B 8].

3.30. Based on Exercise 3.29 and assuming that $k(t) \in \mathcal{K}$ (Definition 3.41 in Sec.3.4), comment on the controllability properties of the system (3.69) for $a \geq 0$ and for $a < 0$. Here $k(t)$ is to be treated as the control variable. Extend these results and those of Ex.3.29 for the nonlinear system

$$(d^2y/dt^2) + a(dy/dt) + k(t)\varphi(y) = 0, \quad t \geq 0 \tag{3.70}$$

with $k(t) \in \mathcal{K}$ and $\varphi(\cdot) \in C$ for both $a \geq 0$ and $a \leq 0$. Based on these, can one infer exponential stability of (3.70) from the exponential stability of (3.69) ? Compare your conclusions with those of Pyatniskii [P 10(a)] and with the comments of Baker and Bergen [B 1].

3.31. In the course of analysis of the stability of motion of the gyroscopic system of stabilization established on a ship which is performing complicated maneuvering, the following perturbation equations are obtained [P 5].

$$\begin{aligned}
&x_1 + (dx_2/dt) - k_1(t)x_3 = 0\\
&(dx_1/dt) - m_1x_2 + k_1(t)x_4 - m_2x_5 = 0\\
&-k_1(t)x_1 + m_1x_3 + (dx_4/dt) + m_3x_4 = 0\\
&k_1(t)x_2 + (dx_3/dt) - (m_4 + k_2t))x_4 = 0\\
&m_5x_2 + (dx_5/dt) + m_3x_4 = 0.
\end{aligned} \tag{3.71}$$

Here $x_1, \cdots, x_5$ are the coordinates of the system, $m_1, \cdots, m_5$ the constant coefficients, $k_1(t)$ and $k_2(t)$ some time functions, determined by the maneuvering of the ship. According to Pirogov [P 5], for the following values of the constant parameters,

$$m_1 = (1.54)10^{-6}\text{sec}^2 \; ; \; m_2 = (0.924)10^{-6}\ \text{sec}^{-2} \; ;$$
$$m_3 = 6(10^{-3})\text{sec}^{-1} \; ; \quad m_4 = 25974 \; ; \; m_5 = 10^{-3}\ \text{sec}^{-1},$$

the following asymptotic stability bounds are obtained for $k_1(t)$ and $k_2(t)$:

$|k_1(t)| \leq (0{\cdot}75)\ 10^{-3}\ \sec^{-1}$; $|k_2(t)| \leq 2$ for all $t \geq 0$. If $k_1(t)$ and $k_2(t)$ are time invariant, the Hurwitz bounds are respectively $(1{\cdot}24)\ 10^{-3}\ \sec^{-1}$ and 2. Check these values and determine whether improved stability conditions can be obtained by applying the results of Sec.3.4. Make suitable assumptions and use the transformation, if necessary, given in [R 2] for conversion of the set of equations (3.71) into the form (2.21)..

CHAPTER 4

EXPONENTIAL STABILITY OF FEEDBACK SYSTEMS IN INTEGRAL FORM

4.1. Introduction. The method adopted in this chapter for analysing the exponential stability of nonlinear time varying feedback systems in integral form,

$$\sigma(t) = f_o(t) - \int_{t_o}^{t} g(t-\tau)k(\tau)\varphi(\sigma(\tau))d\tau , \quad t \geq t_o \geq 0 \tag{2.6}$$

under Assumption S2 of Sec.2.32 (Problem 2.41 of Chapter 2), constitutes a significant departure from the Lyapunov tradition. We also consider on the same lines the linear time varying feedback system described by

$$\sigma(t) = f_o(t) - \int_{t_o}^{t} g(t-\tau)k(\tau)\sigma(\tau)d\tau, \quad t \geq t_o \geq 0 . \tag{2.22}$$

As explained in Sec.3.2 , the early stability criteria, of the pioneering Russian school even for the time invariant nonlinear system (2.24), derived from an application of Lyapunov's classic theorem [L 10] , did not have sufficient algebraic simplicity. Further, there remains the problem of constructing a suitable Lyapunov function candidate.

Popov's criterion, expressed in terms of a simple frequency domain inequality, (Theorem 3.52 with $k(t) \equiv$ a constant) dispensed with the construction of a Lyapunov function candidate. The innovation rekindled interest in the derivation of algebraic frequency domain criteria for the stability of other nonlinear systems, both time invariant and time varying.

The method of Popov consists in obtaining an integral inequality for the solutions of (2.6) and (2.22) of the form

$$\eta_1\sigma^2(T) + \eta_2\int_0^T \sigma^2(t)dt \leq \eta_3\sigma^2(0) \tag{4.1}$$

for some positive constants η_1, η_2 and η_3, and for all $T > 0$.
To this end, suitable quadratic functionals are chosen for the linear and nonlinear systems and evaluated along their trajectories (solutions). Fourier transformation is used to evaluate the upper bound of the functionals.

Essentially, the problem of exponential stability of (2.6) and (2.22) amounts to the following : Find conditions on $G(j\omega)$ and $k(t)\varphi(\cdot)$ as regards the nonlinear system (2.6), and on $G(j\omega)$ and $k(t)$ as regards the linear system (2.22) such that $|\sigma(t)| \leq \gamma_o \psi(\sigma(t_o)) \exp(-\gamma_1(t-t_o))$ for some positive constants γ_o and γ_1 and for all $t \geq t_o \geq 0$. At the present stage of development of stability theory, these conditions seem to be derivable only under Assumption S2 (specialized to the linear time invariant system (2.16)) of Sec.2.32. Further, in the interest of simplicity, generality of the results possible through using the Popov method is sacrificed by assuming that $f_o(t)$ and $g(t)$ have all the quantitative properties of $\underline{c}'\exp(A_o(t-t_o))\underline{x}_o$ and $\underline{c}'\exp(A_o t)\underline{b}$ respectively which are used to define them in Sec.2.2.

Motivated by the form of the Popov criterion, stability conditions are expressed in terms of certain multiplier functions (see Sec.3.2 and Lemma 3.22 for the background). In contrast with the results of Chapter 3, the following are the improvements in the present chapter : (i) Use of noncausal multiplier functions for both linear and nonlinear systems with $\varphi(\cdot) \in C_m$ and C_{mo} (Sec. 4.3 and Sec.4.4);

(ii) Incorporation of the period T_o of a periodic time varying gain $k(t)$ in the stability conditions for both linear and nonlinear systems with $\varphi(\cdot) \in C_m$ and C_{mo} (Sec.4.3 and Sec.4.4) ;

(iii) Removal of the usual restriction on $\theta(t)$, the normalized rate of variation, in the case of the nonlinear system with $\varphi(\cdot) \in C_{mo}$ (Sec.4.4) ;

(iv) Simultaneous upper and lower global bounds on $\theta(t)$, a trade-off between the two being possible and desirable in applications (Sec.4.4 and Sec.4.5).

A comparison with the literature (Sec.4.5(b)) shows that the results are perhaps the best available. Still the basic problems outlined in Sec.3.7 remain unsolved.

4.2. Preliminaries. (a) For definitions of causal, anticausal and noncausal functions, see Sec.3.2. However, it is not necessary here to assume that Z(s), the chosen multiplier function, is rational.

Recall the following definitions used in Sec.3.5 :

$$\Phi(\sigma) = \int_0^{\sigma} \varphi(\sigma)d\sigma > 0 \quad \text{for} \quad \sigma \neq 0 \; ; \quad \delta(\sigma) = (\Phi(\sigma)/\varphi(\sigma)\sigma) \; ;$$

$$\delta_s = \sup_{\sigma} \delta(\sigma) \; ; \quad \text{and} \quad \delta_i = \inf_{\sigma} \delta(\sigma) \tag{4.2}$$

$\delta_D(t-T)$ denotes an impulse occuring at $t = T$.

Further, for convenience in manipulations, the subscript o to f in (2.6) and (2.22) is omitted in what follows.

Based on Lemma 3.22 and Remark 3.24, the main idea in establishing stability conditions for (2.6) and (2.22) under Assumption S2 of Sec.2.32 is to find a complex function of frequency $M(j\omega)$ such that with the arg $\{M(j\omega)\}$ lying in the band (3.5), (i) its introduction into the feedback system (Exercise 2.3) does not lead to instability ; and (ii) the cascade connection of the system representing $M(j\omega)$ and $k(t)\varphi(\cdot)$ is passive.

While it has not yet been possible, in the Lyapunov framework of Chapter 3, to employ an arbitrary (but rational) function $M(j\omega)$ with its argument in the band (3.5), Popov's method does enable us to deal with such functions. The key to the superiority of the Popov method is the following : It is well known from circuit theory that if $Z(s)$ is a positive real function, and $z(t)$ is its inverse transform, then for any function $f(t)$ in the class of continuous real valued functions, the functional

$$\rho_o(T) = \int_0^T f(t)\left(\int_0^t z(t-\tau)f(\tau)d\tau\right)dt \geq 0 \tag{4.3}$$

for all $T > 0$. Invoking the Parseval theorem on Fourier transforms (Lemma 4.41) V.M.Popov showed that if $z(t) \in L_1 \cap L_2(0,\infty)$, then

$$\operatorname{Re} Z(j\omega) \geq 0 \quad \text{for all real } \omega \tag{4.4}$$

alone implies that (4.3) is satisfied.

An extension of this result to noncausal functions leads to the next result.

Lemma 4.21. If $Z_{nc}(j\omega)$ is a noncausal function and $z_{nc}(t)$ is its inverse Fourier transform, then the functional

$$\rho_1(T) = \int_0^T f(t)\left(\int_{-\infty}^{\infty} z_{nc}(t-\tau)f(\tau)d\tau\right)dt \qquad (4.5)$$

is nonnegative for any f(t) in the class of continuous real valued functions if (i) $z_{nc}(t) \in L_1 \cap L_2(-\infty,\infty)$, and (ii) Re $Z_{nc}(j\omega) \geq 0$ for all real ω.

This lemma is a special case of Lemma 4.37 below.

As the Parseval theorem of the theory of Fourier transforms is used in deriving stability criteria, the existence of Fourier transforms of whatever integrals we consider involving $\sigma(t)$ and f(t) of (2.6) and (2.22) should be guaranteed. Since the problem itself is the determination of conditions under which $\sigma(t)$ has suitable qualitative (i.e., stability) properties, no a priori assumptions about $\sigma(t)$ for $t \in [0,\infty)$ can be made. It goes to the credit of V.M.Popov [P 8(a)] for having introduced the truncated function of time into stability analysis. The same technique is used in the Zames-Sandberg approach.

The main assumption, in the use of truncated function in stability analysis is that the solutions of (2.6) and (2.22) do not become unbounded for finite values of time. With this assumption, it is then shown that a certain inequality, (namely,(4.1)) concerning the solutions, holds irrespective of the interval of time, and hence the solutions have the required stability property.

However, it should be pointed out that a stability result with any claim to generality should be capable of predicting the finite time or infinte time properties of the system solutions. That is to say, no assumptions about even the finite time solution behaviour need be made , and whatever solutions exist have then the property expressed in the stability conditions. Results along these lines do not seem to be available..

4.3. Basic lemmas on nonnegative quadratic functionals.

Let

$$\begin{aligned} \sigma_T(t) &= \sigma(t) \quad \text{for } 0 \leq t \leq T \\ &= 0 \quad \text{for } t > T \end{aligned} \qquad (4.6)$$

$$\begin{aligned} \varphi_T(\sigma(t)) &= \varphi(t) \text{ for } 0 \leq t \leq T \\ &= 0 \text{ for } t > T \end{aligned} \qquad (4.7)$$

$$f_T(t) = \sigma(t) \text{ for } 0 \leq t \leq T$$
$$= \int_0^T g(t-\tau)k(\tau)\varphi(\sigma(\tau))d\tau \quad \text{for } t > T \tag{4.8}$$

While analysing the linear system (2.22), (4.8) is changed to

$$f_T(t) = f(t) \quad \text{for } 0 \leq t \leq T$$
$$= \int_0^T g(t-\tau)k(\tau)\sigma(\tau)d\tau \quad \text{for } t > T \tag{4.9}$$

In establishing the conditions for the nonnegativity of quadratic functionals containing $k(t)$ in the integrand, the following lemma, known as the second mean value theorem in calculus, is found to be useful.

Lemma 4.31. If the bounded function $\mu(t)$ is monotone and nonincreasing in the interval (t_1,t_2) and everywhere > 0, and if $\lambda\ (t)$ is summable (whether bounded or not) in (t_1,t_2), then

$$\int_{t_1}^{t_2} \mu(t)\lambda(t)dt = \mu(t_1)\int_{t_1}^{t_3} \lambda(t)dt$$

where t_3 is some number such that $t_1 \leq t_3 \leq t_2$. Also if $\mu(t)$ is monotone and nondiminishing in (t_1,t_2) and everywhere > 0, then

$$\int_{t_1}^{t_2} \mu(t)\lambda(t)dt = \mu(t_2)\int_{t_3}^{t_2} \lambda(t)dt$$

where t_3 is some number such that $t_1 \leq t_3 \leq t_2$.

Proof. See Hobson [H 8,p 618] .

We now take up the problem of establishing conditions under which the cascade connection of the time varying nonlinear function, $k(t)\varphi(\cdot)$ and a linear time invariant system (See Exercise 2.3), represented by $M(j\omega)$,which may be causal, anticausal or noncausal, is passive. The linear time invariant system, as indicated above in Sec.4.2, should not make the enlarged system unstable , and has to satisfy the inequality Re $M(j\omega) \geq 0$ for all real ω . The lemmas established below are needed in the proof of the stability theorems. Table 4.1 lists, for convenient reference, the lemmas of the present section and corresponding theorems of Sec.4.4. The other lemmas needed in the proofs of the theorems are Lemmas 4.41 and 4.42 and a result of the type given in Appendix 4.4.

Table 4.1

No.	Lemma of Sec.4.3.	Theorem of Sec.4.4 (Table 4.2) $\varphi(\cdot), k(t)$.	Comments on the Theorem
1	4.36	4.41, C_m	Possibly the most general result known for the system (2.6) considered here. Generalizes Theorem 3.51 (Chapter 3).
2	4.32	4.42, C	Generalization of Popov's criterion. For β arbitrarily small, equivalent to the circle criterion.
3	4.33 (Corollary 4.332)	4.43, C_m	Equivalent to the Zames criterion [Z 3(c)]
4	4.35 (Corollary 4.351)	4.44, C_m	The counterpart of Theorem 4.43.
5	4.35 (Corollary 4.352)	4.45, C_m	Combination of Theorems 4.43 and 4.44.
6	4.35 (Corollary 4.353)	4.46,linear	Generalizes the result of Brockett and Forys [B 17]
7.	4.37,	4.47,linear	See Sundaresan and Thathachar [S 14(c)] for a parallel result.
8	4.36 (Exercise 4.5)	4.48, C_{mo}	See item No.1 above.
9	4.38 and Appendix 4.3.	4.49,C_{mo} k(t) periodic	Possibly the most general result known for the problem considered here.No counterpart using the Lyapunov framework is known.
9	4.39	4.50,linear k(t)periodic	See Willems [W3(a)] for a parallel result using a different approach.

4.31. Positive real multiplier functions. We state a lemma, similar in form to Lemma 3.51, concerning $\varphi(\cdot) \in C$. This sets the pattern for other lemmas.

The functionals under consideration here and in Sec.4.32 are said to be nonnegative if they are nonnegative for all $\sigma_T(t)$ appearing in the integrands.

Lemma 4.32. The functional

$$I_{no} = \int_0^T k(t)\varphi_T(\sigma(t))(\alpha\sigma_T(t) + \beta\, d\sigma_T(t)/dt)dt + \beta\, k(0)\Phi_T(\sigma(0)) \qquad (4.10)$$

where $\varphi(\cdot) \in C$, $\alpha > 0$ and $\beta \geq 0$, is nonnegative if $k(t)\,\exp(-\gamma_o t)$ is nonincreasing for some nonnegative constant $\gamma_o \leq (\alpha/\beta\delta_s)$.

Proof. $I_{no} = \int_0^T k(t)\exp(-\gamma_o t)\exp(\gamma_o t)\varphi_T(\sigma(t))(\alpha\sigma_T(t) + \beta d\sigma_T(t)/dt)dt$

$$+ \beta\, k(0)\, \Phi_T\,(\sigma(0)).$$

Invoke Lemma 4.31 and note that $\Phi_T(\sigma(t)) = \Phi_T(\sigma_T(t))$ to rewrite I_{no} as

$$I_{no} = k(0)\int_0^{T_1} \exp(\gamma_o t)\varphi_T(\sigma(t))(\alpha\sigma_T(t) + \beta\, d\sigma_T(t)/dt)dt + \beta k(0)\Phi_T(\sigma(0)).$$

where T_1 is some number such that $0 \leq T_1 \leq T$. Integrate by parts and perform minor manipulations to get

$$I_{no} = \alpha k(0)\int_0^{T_1} \exp(\gamma_o t)\varphi_T(\sigma(t))\sigma_T(t)dt$$

$$-\beta k(0)\gamma_o \int_0^{T_1} \exp(\gamma_o t)\,\Phi_T(\sigma(t))dt + \beta k(0)\exp(\gamma_o T_1)\,\Phi_T(\sigma(T_1))$$

But $\Phi_T(\sigma(t)) \leq \delta_s\varphi_T(\sigma(t))\sigma_T(t)$. Hence

$$I_{no} \geq (\alpha-\beta\gamma_o\delta_s)k(0)\int_0^{T_1} \exp(\gamma_o t)\varphi_T(\sigma(t))\sigma_T(t)dt + \beta k(0)\exp(\gamma_o T_1)\Phi_T(\sigma(T_1))$$

from which we conclude that I_{no} is nonnegative if $\gamma_o \leq (\alpha/\beta\delta_s)$.

Remark 4.31. Recognize that the integrand in the integral term of (4.10) is the input-output product of a cascaded combination of the system represented by $1/(\alpha+\beta s)$ and the nonlinear component $\varphi(\cdot) \in C$. Lemma 4.32 states the conditions under which this cascaded combination is passive. No other choice of the multiplier function Z(s) is known for which the cascaded combination of the system represented by Z(s) (or $Z^{-1}(s)$) and the nonlinear component $\varphi(\cdot)\in C$ is passive. Extension of the class of multiplier functions necessitates a reduction in the generality of $\varphi(\cdot)$. That is to say, more restrictions like monotonicity and odd monotonicity are to be imposed on $\varphi(\cdot)$ when we choose other multiplier functions.

Lemma 4.33. Let $Z_{1i}(s) = (s+\nu_i\mu_i)/(s+\mu_i) = 1+(\nu_i-1)\mu_i/(s+\mu_i)$ where $\mu_i > 0$ and $0 \leq \nu_i < 1$. Further let

$$(\mathcal{Z}_{1i}\sigma_T)(t) = \sigma_T(t)+(\nu_i-1)\mu_i\int_0^t \exp(-\mu_i(t-\tau))\sigma_T(\tau)d\tau. \tag{4.11}$$

Then the functional

$$I_{ni}=\int_0^T k(t)\varphi_T(\sigma(t))\ (\mathcal{Z}_{1i}\sigma_T)(t)dt \tag{4.12}$$

is nonnegative for $\varphi(\cdot)\in C_m$ if $k(t)\exp(-\gamma_{1i}t)$ is nonincreasing for some nonnegative $\gamma_{1i} \leq (\nu_i\mu_i/\delta_s)$.

Proof. See Appendix 4.1.

Corollary 4.331. Let $Z_1(s) = \sum_i \eta_i(s+\nu_i\mu_i)/(s+\mu_i)$ with $\mu_i > 0$; $\eta_i, \nu_i \geq 0$ and $0 \leq \nu_i < 1$ for all i. Further, let

$$(\mathcal{Z}_1\sigma_T)(t) = \sum_i \eta_i(\sigma_T(t)+(\nu_i-1)\mu_i\int_0^t \exp(-\mu_i(t-\tau))\sigma_T(\tau)d\tau) \tag{4.13}$$

Then the functional

$$I_{n1} = \int_0^T k(t)\varphi_T(\sigma(t))(\mathcal{Z}_1\sigma_T)(t)dt \tag{4.14}$$

is nonnegative for $\varphi(\cdot)\in C_m$ if $k(t)\exp(-\gamma_1 t)$ is nonincreasing for some nonnegative $\gamma_1 \leq \min_i(\nu_i\mu_i/\delta_s)$.

Corollary 4.332. Let $Z_{10}(s) = \alpha+\beta s+\sum_i \eta_i(s+\nu_i\mu_i)/(s+\mu_i)$ with α and $\mu_i > 0$; $\beta, \eta_i, \nu_i \geq 0$ and $0 \leq \nu_i < 1$ for all i. Further let $(\mathcal{Z}_1\sigma_T)(t)$ be defined by (4.13). Then the functional

$$I_{n10}=\int_0^T k(t)\varphi_T(\sigma(t))(\alpha\sigma_T(t)+\beta(d\sigma_T(t)/dt)+(\mathcal{Z}_1\sigma_T)(t))dt+\beta k(0)\Phi_T(\sigma(0)) \tag{4.15}$$

is nonnegative for $\varphi(\cdot)\in C_m$ if $k(t)\exp(-\gamma_{10}\, t)$ is nonincreasing for some nonnegative $\gamma_{10}\leq \min_i(\alpha/\beta,\ \nu_i\mu_i)/\delta_s$.

For functionals involving $\varphi(\cdot)\in C_{mo}$, see Exercise 4.1.

Suppose $\varphi(\sigma)\equiv \sigma$, then Lemmas 4.32 and 4.33 hold with $\delta_s = (1/2)$. In fact by an independent treatment it is possible to improve upon this special result.

Lemma 4.34. Let $Z_{\ell 1}(s) = \alpha+\beta s+ \sum_i \eta_i(s+\nu_i\mu_i)/(s+\mu_i)$ with α and $\mu_i > 0$; β, η_i and $\nu_i \geq 0$ for all i. Further, let $(z_1\sigma_T)(t)$ be defined by (4.13). Then the functional

$$I_{\ell 1} = \int_0^T k(t)\sigma_T(t)(\alpha\sigma_T(t)+\beta(d\sigma_T(t)/dt)+ (z_1\sigma_T)(t))dt+\beta k(0)\sigma_T^2(0)/2$$

is nonnegative if $k(t)\exp(-\gamma_{\ell 1}(t))$ is nonincreasing for some nonnegative $\gamma_{\ell 1}\leq 2\min_i(\alpha/\beta,\ \nu_i\mu_i,\mu_i)$.

Remark 4.32. Note that no restriction on ν_i in the multiplier function $Z_{\ell 1}(s)$ is imposed. The multiplier function corresponds to the one used by Brockett and Forys [B 17] for the linear time varying system (2.21).

The multiplier functions considered above are positive real and rational. Hence the stability results obtained from the use of the corresponding quadratic functionals may not be more general than those of Chapter 3.

4.32. Anticausal and noncausal multiplier functions.

The counterpart of the causal function of Lemma 4.33 is obtained as follows : Let

$$z_a(t) = \sum_i -\eta_i \exp(\mu_i t) \quad \text{for } t\leq 0$$
$$= 0 \quad \text{for } t> 0 \tag{4.16}$$

where $\mu_i > 0$ and $\eta_i \geq 0$ for all i. The Fourier transform of $z_a(t)$ is equal to $\sum_i -\eta_i/(\mu_i-j\omega)$. Hence, the Laplace inverse of $\sum_i -\eta_i/(\mu_i-s)$, interpreted as the transform of an impulse response which does not vanish for negative time, is given by (4.16). Further, let $Z^a_{ni}(s) = ((\mu_i-\eta_i)-s)/(\mu_i-s)$ with the inverse transform $z^a_{ni}(t) = \delta_D(t) - \eta_i \exp(\mu_i t)$ for $t\leq 0$

$$= 0 \quad \text{for } t> 0. \tag{4.17}$$

Define

$$(\mathcal{Z}^a_{ni}\sigma_T)(t) = \sigma_T(t)-\eta_i \int_t^{\infty} \exp(\mu_i(t-\tau))\sigma_T(\tau)d\tau. \tag{4.18}$$

Lemma 4.35. With $(\mathcal{Z}^a_{ni}\sigma_T)(t)$ defined by (4.18) and $\mu_i > \eta_i$, the functional

$$I^a_{ni} = \int_0^T k(t)\varphi_T(\sigma(t))(\mathcal{Z}^a_{ni}\sigma_T)(t)dt \tag{4.19}$$

is nonnegative for $\varphi(\cdot)\in C_m$ if $k(t)\exp(\gamma_{ai}t)$ is nondecreasing for some nonnegative $\gamma_{ai} \leq (\mu_i-\eta_i)/\delta_s$.

Proof. In (4.18), let $\sigma_1 = \sigma_T(t)$ and

$$\sigma_1-\sigma_2 = \sigma_T(t)-\eta_i \int_t^{\infty} \exp(\mu_i(t-\tau))\sigma_T(\tau)dt$$

from which

$$\sigma_2 = \eta_i \int_t^{\infty} \exp(\mu_i(t-\tau))\sigma_T(\tau)d\tau \tag{4.20}$$

and

$$(d\sigma_2/dt) = \mu_i\sigma_2 - \eta_i\sigma_1$$

or

$$\sigma_1 = (\mu_i\sigma_2 - (d\sigma_2/dt))/\eta_i \tag{4.21}$$

Add to and subtract from (4.19) the expression $\varphi_T(\sigma_2)(\sigma_1-\sigma_2)$ to get

$$I^a_{ni} = \int_0^T k(t)(\varphi_T(\sigma_2)-\varphi_T(\sigma_1))(\sigma_1-\sigma_2)dt + \int_0^T k(t)\varphi_T(\sigma_2)(\sigma_1-\sigma_2)dt \tag{4.22}$$

Because $\varphi(\cdot)\in C_m$, the first integral of (4.22) is nonnegative. The second integral of (4.22), on using (4.21) and defining

$$\sigma_3 = ((\mu_i-\eta_i)\sigma_2-(d\sigma_2/dt))/\eta_i,$$

gives

$$\int_0^T k(t)\varphi_T(\sigma_2)(\sigma_1-\sigma_2)dt = \int_0^T k(t)\varphi_T(\sigma_2)\sigma_3 dt \tag{4.23}$$

According to the hypothesis of the lemma, $k(t)\exp(\gamma_{ai}t)$ is nondecreasing for some nonnegative γ_{ai}. Invoke Lemma 4.31 to rewrite (4.23) as

$$\int_0^T k(t)\varphi_T(\sigma_2)(\sigma_1-\sigma_2)dt = k(T)\exp(\gamma_{ai}T) \int_{T_1}^T \exp(-\gamma_{ai}t)\varphi_T(\sigma_2)\sigma_3 dt \tag{4.24}$$

where T_1 is some number such that $0 \leq T_1 \leq T$.

Equation (4.24) contains $\varphi_T(\sigma_2)(d\sigma_2/dt)$ in its integrand. Hence consider the integral

$$\int_{T_1}^{T} \exp(-\gamma_{ai}t)\varphi_T(\sigma_2)(d\sigma_2/dt)dt = \exp(-\gamma_{ai}t)\Phi_T(\sigma_2)\Big]_{T_1}^{T}$$

$$+ \gamma_{ai}\int_{T_1}^{T} \exp(-\gamma_{ai}t)\Phi_T(\sigma_2)dt \qquad (4.25)$$

But σ_2 defined by (4.20) is zero for $t = T$. Hence from (4.25), we have

$$\int_{T_1}^{T} \exp(-\gamma_{ai}t)\varphi_T(\sigma_2)(d\sigma_2/dt)dt = -\exp(\gamma_{ai}T_1)\Phi_T(\sigma_2(T_1))$$

$$+ \gamma_{ai}\int_{T_1}^{T} \exp(-\gamma_{ai}t)\Phi_T(\sigma_2)dt, \qquad (4.26)$$

which on substitution in (4.24) yields

$$\int_0^T k(t)\varphi_T(\sigma_2)(\sigma_1-\sigma_2)dt = (k(T)\exp(\gamma_{ai}T)/\eta_i)\int_{T_1}^{T} \exp(-\gamma_{ai}t)\Big\{(\mu_i-\eta_i)\sigma_2\varphi_T(\sigma_2)$$

$$-\gamma_{ai}\Phi_T(\sigma_2)\Big\}dt + (k(T)\exp(\gamma_{ai}T)/\eta_i)\exp(\gamma_{ai}T_1)\Phi(\sigma_2(T_1)) \qquad (4.27)$$

Hence the expression on the right hand side of (4.27) is nonnegative if $\mu_i > \eta_i$ and $\gamma_{ai} \leq (\mu_i-\eta_i)/\delta_s$. The lemma is proved.

Corollary 4.351. With $(\mathbf{z}^a_{n1}\sigma_T)(t)$ defined by

$$(\mathbf{z}^a_{n1}\sigma_T)(t) = \sigma_T(t) - \sum_i \eta'_i \int_t^{\infty} \exp(\mu'_i(t-\tau))\sigma_T(\tau)d\tau \qquad (4.28)$$

where $\eta'_i \geq 0$, $\mu'_i > 0$ and $\mu'_i > \eta'_i$ for all i, the functional

$$I^a_{n1} = \int_0^T k(t)\varphi_T(\sigma(t))(\mathbf{z}^a_{n1}\sigma_T)(t)dt \qquad (4.29)$$

is nonnegative for $\varphi(\cdot) \in C_m$ if $k(t)\exp(\gamma_{a1}t)$ is nondecreasing for some nonnegative $\gamma_{a1} \leq \min_i(\mu'_i-\eta'_i)/\delta_s$.

Corollary 4.352. With $(\mathbf{z}_1\sigma_T)(t)$ defined by (4.13) and $(\mathbf{z}^a_{n1}\sigma_T)(t)$ defined by (4.28), the functional

$$I^a_{nc} = \int_0^T k(t)\varphi_T(\sigma(t))\Big\{\alpha\sigma_T(t)+\beta(d\sigma_T(t)/dt)+(\mathbf{z}_1\sigma_T)(t)+(\mathbf{z}^a_{n1}\sigma_T)(t)\Big\}dt$$

where $\alpha > 0$ and $\beta \geq 0$, is nonnegative for $\varphi(\cdot) \in C_m$ if (i) $k(t)\exp(-\gamma_{20}t)$ is nonincreasing ; (ii) $k(t)\exp(\gamma_{a2}t)$ nondecreasing for some nonnegative constants

$$\gamma_{20} \leq \min_i(\alpha/\beta, \nu_i\mu_i)/\delta_s, \quad \gamma_{a2} \leq \min_i(\mu_i'-\eta_i')/\delta_s \tag{4.30}$$

where $0 \leq \nu_i < 1$ and $\mu_i' > \eta_i'$.

Corollary 4.353. If, in Corollary 4.352, $\varphi(\sigma) \equiv \sigma$, then a result better than merely replacing δ_s by (1/2) is possible. The corollary holds with

$$\begin{aligned} &\gamma_{30} \leq 2 \min_i(\alpha/\beta, \nu_i\mu_i) \text{ for } \nu_i \geq 0; \\ &\gamma_{a3} \leq 2 \min_i(\mu_i'-\eta_i') \text{ for } \mu_i' > \eta_i' \geq 0; \\ \text{or} \quad &\gamma_{a1} \leq 2 \min_i \mu_i' \quad \text{for } \eta_i' < 0, \end{aligned} \tag{4.31}$$

replacing γ_{20} and γ_{a2} respectively. (Check these conclusions).

See Exercise 4.2 for a result concerning $\varphi(\cdot) \in C_{mo}$.

Remark 4.33. Lemmas 4.34 and 4.35 (and corollaries) contain results on nonnegativity of a quadratic functional for only specific forms of a rational multiplier function. It is possible to generalize these by considering, for instance, biquadratic functions (Exercise 4.3), and also to derive similar results for $\varphi(\cdot) \in$ class of power law functions as introduced in Exercise 3.13 (Exercise 4.4). Instead of dealing with specific forms of multiplier functions, we now give results for multiplier functions which are not necessarily rational.

Lemma 4.36. Let $z_c(\cdot)$ be a real valued function on $[0,\infty)$ and $z_c(t) = 0$ for $t < 0$ and there exist positive constants η_1 and η_2 such that $|z_c(t)| \leq \eta_1 \exp(-\eta_2 t)$ for all $t \geq 0$. Further, let $z_{ac}(\cdot)$ be a real valued function on $(-\infty,0]$ and $z_{ac}(t) = 0$ for $t > 0$ and there exist positive constants η_1' and η_2' such that $|z_{ac}(t)| \leq \eta_1' \exp(\eta_2' t)$ for all $t \leq 0$. Also let z_{ci} and z_{aci} be real constants for $i = 1,2,3,\cdots$, such that $\sum_i (|z_{ci}| + |z_{aci}|) < \infty$. Define

$$\begin{aligned}(Z_g\sigma_T)(t) = &\sum_i z_{ci}\sigma_T(t-\tau_i) + \sum_{i'} z_{aci}\,\sigma_T(t+\tau_{i'}') + \\ &+ \int_0^\infty z_c(\tau)\sigma_T(t-\tau)d\tau + \int_{-\infty}^0 z_{ac}(\tau)\sigma_T(t-\tau)d\tau + \alpha\sigma_T(t)\end{aligned} \tag{4.32}$$

where $\alpha > 0$ and $0 < \tau_1$, $\tau_1' < \tau_2$, τ_2' $\cdots$. The multiplier function corresponding to the operator Z_g is

$$Z_g(j\omega) = \sum_i z_{ci} \exp(-ij\omega\tau_i) + \sum_{i'} z_{aci'} \exp(-i'j\omega\tau_{i'}') + Z_c(j\omega)$$

$$+ Z_{ac}(j\omega) + \alpha. \tag{4.33}$$

Then the functional

$$I_{ng} = \int_0^T k(t)\varphi_T(\sigma(t))(Z_g\sigma_T)(t)dt \tag{4.34}$$

is nonnegative for $\varphi(\cdot) \in C_m$ if

(i) z_{ci}, z_{aci} are nonpositive for all $i = 1,2,3,\cdots$; $z_c(\cdot)$, $z_{ac}(\cdot)$ are also nonpositive in their intervals of definition ;

(ii)
$$\sum_i |z_{ci}| \exp(\gamma_c\tau_i) + \sum_{i'} z_{aci'} \exp(\gamma_{ac}\tau_{i'}') + \int_0^\infty |z_c(t)| \exp(\gamma_c t)dt$$

$$+ \int_{-\infty}^0 |z_{ac}(t)| \exp(-\gamma_{ac} t)dt \leq \alpha/(1+\delta_s-\delta_i) \tag{4.35}$$

for some nonnegative constants γ_c and γ_{ac} ; and

(iii) $k(t)\exp(-\gamma_c t)$ is nonincreasing, $k(t)\exp(\gamma_{ac} t)$ is nondecreasing.

Proof. See Appendix 4.2.

Remark 4.34. Lemma 4.36 seems to be the most general result concerning the passivity of a nonlinear time varying gain and a linear time invariant noncausal system. It is a generalization of the result of O'Shea [O 2] derived for a nonlinear time invariant gain. Observe that the result of O'Shea does not contain γ_c, γ_{ac} and, as a consequence, $(\delta_s - \delta_i)$ in the integral inequality corresponding to (4.35). When applied to the time invariant nonlinear problem, γ_c and γ_{ac} in hypothesis (ii) of Lemma 4.36 can be chosen arbitrarily small. But then the result of O'Shea [O 2] cannot be obtained as a special case.

A trade-off between γ_c and γ_{ac} in hypothesis (ii) of Lemma 4.36 is permissible with the result that the behaviour of $k(t)$, as evident from hypothesis (iii), is also allowed some flexibility. For $\gamma_{ac} = -\infty$, choice of $z_{aci} = 0$ for all i and $z_{ac}(t) = 0$ for all $t \leq 0$ in Lemma 4.36 gives Lemma 3.53 with the improvement that the corresponding multiplier function $(\alpha + \sum_i z_{ci}\exp(-s\tau_i) + Z_c(s))$ need not be positive real. For $\varphi(\cdot) \in C_{mo}$, hypothesis (i) of the lemma can be bypassed (Exercise 4.5)

Remark 4.35. When $\varphi(\sigma) \equiv \sigma$, Lemma 4.36 holds with the right hand side of (4.34) replaced by 1, because $\delta_s = \delta_i = (1/2)$. As in the case of Lemmas 4.33 and 4.35, specialization of a result on the nonlinear system to that of a linear system does not always give the best result possible for the latter. By treating the linear system problem separately, we get the following result.

Lemma 4.37. Let $z_c(t)$, $z_{ac}(t)$, z_{ci} and z_{aci} be as defined in Lemma 4.36. Define

$$(\mathcal{Z}_{\ell g}\sigma_T)(t) = \sum_i z_{ci}\sigma_T(t-\tau_i) + \sum_{i'} z_{aci'}\sigma_T(t+\tau'_{i'}) + \int_0^{\infty} z_c(\tau)\sigma_T(t-\tau)d\tau + \int_{-\infty}^{0} z_{ac}(\tau)\sigma_T(t-\tau)d\tau \tag{4.36}$$

where $0 < \tau_1, \tau'_1 < \tau_2, \tau'_2 < \cdots$. Then the functional

$$I_{\ell g} = \int_0^T k(t)\sigma_T(t)(\mathcal{Z}_{\ell g}\sigma_T)(t)dt \tag{4.37}$$

is nonnegative for all $\sigma_T(t)$ if

(i) $$\mathrm{Re}\Big\{\sum_i z_{ci}\exp((-j\omega - \gamma_{c\ell})\tau_i) + Z_c(j\omega - \gamma_{c\ell})\Big\} \geq 0 ,$$
$$\mathrm{Re}\Big\{\sum_i z_{aci}\exp((j\omega + \gamma_{ac\ell})\tau'_i) + Z_{ac}(j\omega + \gamma_{ac\ell})\Big\} \geq 0 \tag{4.38}$$

for some nonnegative constants $\gamma_{c\ell}$ and $\gamma_{ac\ell}$ and for all real ω ;

(ii) $k(t)\exp(-\gamma_{c\ell}\, t)$ is nonincreasing, $k(t)\exp(\gamma_{ac\ell}\, t)$ is nondecreasing.

Proof. The proof involves the use of Lemma 4.31 and the Parseval theorem (see Lemma 4.41 below) in the theory of Fourier transforms, and is suggested as an exercise (Exercise 4.6).

Corollary 4.361. Let $(\mathcal{Z}_g\sigma_T)(t)$ be as defined by (4.32). Then the functional

$$I'_{ng} = \int_0^T k(t)\varphi_T(\sigma(t))\Big\{(1-\alpha)\sigma_T(t) + \beta(d\sigma_T(t)/dt) + (\mathcal{Z}_g\sigma_T)(t)\Big\}dt + \beta k(0)\Phi_T(\sigma(0))$$

where $0<\alpha<1$ and $\beta\geq 0$, is nonnegative for $\varphi(\cdot)\in C_m$ if hypotheses (i) and (ii) of Lemma 4.36 hold, and in hypothesis (iii), γ_c is replaced by min $((1-\alpha)/\beta\delta_s, \gamma_c)$.

Corollary 4.371. Let $(Z_{\ell g}\sigma_T)(t)$ be as defined by (4.36). Then the functional

$$I'_{\ell g} = \int_0^T k(t)\sigma_T(t)\{\alpha\sigma_T(t)+\beta(d\sigma_T(t)/dt)+(Z_{\ell g}\sigma_T)(t)\}dt + (\beta k(0)/2)\sigma_T^2(0)$$

where $\alpha>0$ and $\beta\geq 0$, is nonnegative for all $\sigma_T(t)$ if hypothesis (i) of Lemma 4.37 holds and in hypothesis (ii), $\gamma_{c\ell}$ is replaced by min $(2\alpha/\beta, \gamma_{c\ell})$.

Remark 4.36. Observe that in Lemma 4.37 no time domain restriction of the type (4.35) on the multiplier function is imposed. When applied to a periodic k(t), both the Lemmas 4.36 and 4.37 involve, for the nonnegativeness of the functionals, a restriction on the rate of variation of k(t). These nonnegative functionals are used in the establishment of stability conditions for the nonlinear and linear systems (2.6) and (2.22) respectively. Hence the stability conditions will contain a constraint on the rate of variation of k(t) even for periodically varying gains. In view of the fact that the criteria of Lyapunov, Borg and Krein [S 9] for a linear second order periodic coefficient system do not involve such constraints, it is natural to attempt nonnegativeness conditions for quadratic functionals (4.34) and (4.37) containing a periodic k(t), without restrictions on the rate of variation of k(t). This we accomplish below (Lemma 4.38).

Remark 4.37. Note that Lemma 4.36 has been established for $\varphi(\cdot)\in C_m$. Suppose it can be shown that this lemma (or its Corollary 4.361) holds for $\varphi(\cdot)\in C$ and $k(t)\equiv$ a constant under suitable additional conditions on the multiplier function. Then we would be having an important lemma which can be used to show that the well known Popov criterion [P 8(a)] is not necessary. Compare this argument with the counterexample-type of argument in Pyatnitskii [P 10(b)].

Lemma 4.38. Let z_{ci} and z_{aci} be real constants for $i = 1,2,3,\cdots$, such that $\sum_i (|z_{ci}| + |z_{aci}|)<\infty$. Define

$$(Z_p\sigma_T)(t) = \sum_i z_{ci}\sigma_T(t-iT_0) + \sum_{i'} z_{aci'}\sigma_T(t+i'T_0) + \sigma_T(t) \qquad (4.39)$$

Then the functional

$$I_{np} = \int_0^T k(t)\varphi_T(\sigma(t))(\mathcal{Z}_p\sigma_T)(t)dt$$

is nonnegative for $\varphi(\cdot) \in C_m$ and $k(t)$ periodic with T_o as the fundamental period, if (i) $z_{ci} = z_{aci}$ and $z_{ci} \leq 0$ for all $i = 1,2,3,\cdots,$; and

(ii) $\sum_i |z_{ci}| \leq (1/2)$.

Proof. See Appendix 4.3 which also contains the result for $\varphi(\cdot) \in C_{mo}$.

Corollary 4.381. Let $(\mathcal{Z}_p\sigma_T)(t)$ be as defined by (4.39). Then the functional

$$I''_{np} = \int_0^T k(t)\varphi_T(\sigma(t)) \left\{\alpha\sigma_T(t) + \beta(d\sigma_T(t)/dt) + (\mathcal{Z}_p\sigma_T)(t)\right\} dt + \beta\, k(0)\Phi_T(\sigma(0))$$

where $\alpha > 0$ and $\beta \geq 0$, is nonnegative for $\varphi(\cdot) \in C_m$, if hypotheses (i) and (ii) of Lemma 4.38 hold, and β is sufficiently small.

Remark 4.38. On applying Lemma 4.38 to the case $\varphi(\sigma) \equiv \sigma$, we do not get the best result possible. In fact, the transition from $\varphi(\cdot) \in C_m$ or C_{mo} to the class of linear functions $\varphi(\sigma) \equiv \sigma$, is abrupt. (See Remark 4.35). In order to provide a gradual transition from the class C_{mo} to the class of linear functions, it is necessary to consider the class of power law functions introduced in Exercise 3.13, and the results for this class can be specialized to those for the class of linear functions..

We now state a result, similar to Lemma 4.37, applicable to $\varphi(\sigma) \equiv \sigma$ and $k(t)$ periodic.

Lemma 4.39. Let z_{ci} and z_{aci} be real constants for $i = 1,2,3,\cdots$, such that $\sum_i (|z_{ci}| + |z_{aci}|) < \infty$. Define

$$(\mathcal{Z}_{\ell p}\sigma_T)(t) = \sum_i z_{ci}\sigma_T(t - iT_o) + \sum_{i'} z_{aci'}\sigma_T(t + i'T_o). \qquad (4.40)$$

Then the functional

$$I_{\ell p} = \int_0^T k(t)\sigma_T(t)(\mathcal{Z}_{\ell p}\sigma_T)(t)dt$$

is nonnegative for $k(t)$ periodic with T_o as the fundamental period, if

$$\mathrm{Re}\left\{\sum_i z_{ci}\exp(-ij\omega T_o) + \sum_{i'} z_{aci'}\exp(i'j\omega T_o)\right\} \geq 0 \tag{4.41}$$

for all real ω (Exercise 4.7).

4.4. Exponential Stability Criteria.

Based on the above preliminaries and the Parseval theorem (Lemma 4.41 below) in the theory of Fourier transforms, conditions on $G(j\omega)$ and $k(t)\varphi(\cdot)$ (or $k(t)$) in terms of a suitably chosen multiplier function can be established for the solutions of (2.6) (or (2.22)) to satisfy an equality of the type (4.1). From this inequality, stability and, under certain extra assumptions, asymptotic (and exponential) stability (Lemma 4.42 below) of the solutions of (2.6) and (2.22) are guaranteed.

Only one major stability result for (2.6) with $\varphi(\cdot) \in C_m$ is stated and proved. Others listed in Table 4.2 can be established along similar lines.

Lemma 4.41 (Parseval's Theorem). Suppose $f_1(\cdot)$ and $f_2(\cdot)$ are real valued functions defined on $[0,\infty)$ and belong to the class of $L_1(0,\infty)$ functions. Let $F_1(j\omega)$ and $F_2(j\omega)$ be their Fourier transforms. Then

$$\int_0^\infty f_1(t)f_2(t)dt = (1/2\pi)\int_{-\infty}^\infty F_1(j\omega)F_2(-j\omega)d\omega .$$

Lemma 4.42. (Popov-Barbalat) [P 8(a)] . Let $y(t)$ map $[0,\infty)$ into the real line, and be differentiable. If y and (dy/dt) are bounded on $[0,\infty)$ and $\int_0^\infty y^2 dt < \infty$, then $\lim_{t\to\infty} y(t) = 0$. Suppose further that $\int_0^\infty \exp(\gamma t)y^2 dt < \infty$ for some constant $\gamma > 0$, then $\lim_{t\to\infty} y(t)\exp(\gamma t/2) = 0$. Hence $y(t)$ is exponentially decreasing.

Proof. For the first part of the proof, see Popov [P 8(a)] or Desoer [D 4(b)]. The second part of the proof is an extension of the first.

4.41. Major Result.

Let $(Z\sigma_T)(t)$ be defined by

$$(Z\sigma_T)(t) = (Z_g\sigma_T)(t) + \beta(d\sigma_T/dt) + (1-\alpha)\sigma_T(t) \tag{4.42}$$

where $0 < \alpha < 1$, and $(Z_g\sigma_T)(t)$ is as given in Lemma 4.36 by (4.32). The multiplier function corresponding to (4.42) is

$$Z(j\omega) = \sum_i z_{ci}\exp(-ij\omega\tau_i) + \sum_{i'} z_{aci'}\exp(i'j\omega\tau'_{i'}) + Z_c(j\omega) + Z_{ac}(j\omega) + 1 + \beta j\omega \qquad (4.43)$$

where $\beta > 0$, $Z_c(j\omega)$ and $Z_{ac}(j\omega)$ are the Fourier transforms of $z_c(\cdot)$ and $z_{ac}(\cdot)$ respectively introduced in Lemma 4.36. Observe that, with a time domain restriction of the type (4.35) on $Z(j\omega)$, we have the inequality $\mathrm{Re}\, Z(j\omega) \geq 0$ for all real ω. (For convenience in the manipulations to follow, $f_o(t)$ of (2.6) is replaced by $f(t)$).

Theorem 4.41. The nonlinear time varying system (2.6) with $\varphi(\cdot) \in C_m$ has solutions with the property

$$\eta_1 \int_0^T \exp(\gamma t)\sigma_T^2(t)dt + \eta_o \sup_{0 \leq t \leq T} \sigma^2(t) \leq R_o + R_1 \sup_{0 \leq t \leq T} |\sigma(t)| \qquad (4.44)$$

for some positive constants η_1, η_o, γ, R_o and R_1 independent of T and for all $T > 0$, if there exists a multiplier function $Z(j\omega)$ of the form (4.43) such that

(i) $\mathrm{Re}\, Z(j\omega)G(j\omega) \geq 0$ for all real ω ;

(ii) hypotheses (i) and (ii) of Lemma 4.36 are satisfied ;

(iii) hypothesis (iii) of Lemma 4.36 is satisfied with γ_c replaced by $\gamma'_c = \min((1-\alpha)/\beta\delta_s, \gamma_c)$.

Proof. With $\sigma_T(t)$, $\varphi_T(\sigma(t))$ and $f_T(t)$ given by (4.6), (4.7) and (4.8) respectively, define

$$\rho_1(T) = \int_0^T \Big\{(1-\alpha)(\sigma_T(t)-f_T(t)) + \beta\, d(\sigma_T(t)-f_T(t))/dt\Big\}\, k(t)\varphi_T(\sigma(t))dt + \int_0^T k(t)\varphi_T(\sigma(t))((z_g\sigma_T)(t)-(z_g f_T)(t))dt \qquad (4.45)$$

for some positive constant $\alpha < 1$. By virtue of the truncation property of the components of the integrand, $\rho_1(T)$ can be rewritten as

$$\rho_1(T) = \int_0^\infty \Big\{(1-\alpha)(\sigma_T(t)-f_T(t)) + \beta\, d(\sigma_T(t)-f_T(t))/dt\Big\}\, k(t)\varphi_T(\sigma(t))dt$$

$$+ \int_0^\infty k(t)\varphi_T(\sigma(t))((Z_g\sigma_T)(t)-(Z_gf_T)(t))dt \qquad (4.46)$$

Let the Fourier transforms of $\sigma_T(t)$, $k(t)\varphi_T(\sigma(t))$ and $f_T(t)$ be respectively, $\Sigma_T(j\omega)$, $\Psi_T(j\omega)$ and $F_T(j\omega)$. By virtue of the truncation of the integrand components, Lemma 4.41 is applicable to (4.46). Note that $\sigma_T(0) = f_T(0)$, from which the Fourier transform of $d(\sigma_T(t)-f_T(t))/dt$ is $j\omega(\Sigma_T(j\omega)-F_T(j\omega))$. Further, from (2.6), and (4.6)-(4.8), $\Sigma_T(j\omega)-F_T(j\omega) = -G(j\omega)\Psi_T(j\omega)$ and the Fourier transform of $d(\sigma_T(t)-f_T(t))/dt$ is $-j\omega\, G(j\omega)\,\Psi_T(j\omega)$. Also the Fourier transform of $(Z_g\sigma_T)(t)-(Z_gf_T)(t)$ is $-Z_g(j\omega)G(j\omega)F_T(j\omega)$. On application of the Parseval theorem to (4.46), we get

$$\rho_1(T) = -(1/2\pi)\int_{-\infty}^{\infty} \mathrm{Re}\ Z\,(j\omega)G(j\omega)\ |\Psi_T(j\omega)|^2\, d\omega.$$

Since $|\Psi_T(j\omega)|^2$ is real, $\rho_1(T) \leq 0$ according to hypothesis (i) of the theorem statement, implying thereby that

$$\int_0^T ((1-\alpha)\sigma_T(t)+\beta\, d\sigma_T(t)/dt)\ k(t)\varphi_T(\sigma(t))dt + \int_0^T k(t)\varphi_T(\sigma(t))(Z_g\sigma_T)(t)dt$$

$$\leq \int_0^T \left\{(1-\alpha)f_T(t)+\beta(df_T(t)/dt) + (Z_gf_T)(t)\right\} k(t)\varphi_T(\sigma(t))dt \qquad (4.47)$$

for some positive constant $\alpha < 1$.

Now consider the left hand side of the inequality (4.47). From Lemma 4.32, the first integral has the lower bound

$$\eta_1 \int_0^{T_1} \exp(\gamma_o t)\sigma_T^2(t)dt + \eta_o\sigma_T^2(T_1) - \beta\, k(0)\ \Phi_T(\sigma(0))$$

for some positive constants η_1 and η_o and for some T_1 such that $0 \leq T_1 \leq T$, if $k(t)\exp(-\gamma_o t)$ is nonincreasing for some nonnegative constant $\gamma_o < (1-\alpha)/\beta\delta_s$. The second integral on the left hand side of (4.47) is nonnegative under hypothesis (ii) which includes the requirement that $k(t)\exp(-\gamma_c t)$ is nonincreasing for some nonnegative γ_c obtained from hypothesis (ii) of Lemma 4.36. Hence with $\gamma_c' > 0$ chosen to be less than $\min((1-\alpha)/\beta\delta_s,\ \gamma_c)$, the left hand side of the inequality (4.47) has under hypotheses (ii) and (iii) the lower bound

$$\eta_1 \int_0^{T_1} \exp(\gamma_c' t)\sigma_T^2(t)dt + \eta_o\sigma_T^2(T_1) - \beta\, k(0)\, \Phi_T(\sigma(0))$$

for some positive constants η_1 and η_o and for some T_1 such that $0 \leq T_1 \leq T$.

The right hand side of the inequality (4.47) has the upper bound

$$R_1 \sup_{0 \leq t \leq T} |\sigma(t)|$$

for some constant R_1 independent of T. (See Appendix 4.4). (This upper bound demonstrates that the anticausal part of the multiplier function does not lead to instability).

Hence under hypotheses (i)-(iii), the solutions of (2.6) satisfy the inequality

$$\eta_1 \int_0^{T_1} \exp(\gamma_c' t)\sigma_T^2(t)dt + \eta_o\sigma_T^2(T_1) \leq R_o + R_1 \sup_{0 \leq t \leq T} |\sigma(t)| \quad (4.48)$$

for some constant R_o independent of T.

Since the inequality (4.48) has been stated for every T, the solutions of (2.6) have the property (4.44). The theorem is proved.

Remark 4.41. Each term on the left hand side of the inequality (4.44) is positive and hence individually less than the right hand side of (4.44). This gives rise to the inequality

$$\eta_o \sup_{0 \leq t \leq T} \sigma^2(t) \leq R_o + R_1 \sup_{0 \leq t \leq T} |\sigma(t)|$$

from which one concludes [P 8(a)] that $|\sigma(t)| \leq R_2$ for some constant R_2 independent of T, for all $t \geq 0$. Hence, invoking inequality (4.48), we find that

$$\eta_1 \int_0^{T} \exp(\gamma_c' t)\sigma_T^2(t)dt \leq R_3 \quad (4.49)$$

where R_3 is a constant independent of T.

Now differentiate (2.6) to get

$$(d\sigma/dt) = (df/dt) - g(0)k(t)\varphi(\sigma(t)) - \int_0^t (\partial g(t-\tau)/\partial t)k(\tau)\varphi(\sigma(\tau))d\tau. \quad (4.50)$$

If $f(t)$ is obtained from the differential equation description of the system, then under Assumption S1 (Sec.2.32,p 30), $|df(t)/dt| \leq$ a constant for all $t \geq 0$. Hence, given that $|\sigma(t)| \leq R_2$ for all $t \geq 0$, we conclude from (4.50) that $|d\sigma/dt|$ is bounded for all $t \geq 0$. Then Lemma 4.42 applied to (4.49) guarantees the exponential stability of the system (2.6).

4.42. Choice of other multiplier functions.

Stability results for other multiplier functions can be obtained in a similar manner. These are summarized in Table 4.2 which is drawn up so as to fit the framework of the statement of Theorem 4.41. See Table 4.1 for a correpondence between the theorems and the preliminary lemmas used in their proofs.

Table 4.2

No.	Theorem	Class of $\varphi(\cdot)$ $k(t)$	Multiplier function	$0<\gamma_c<$	$0<\gamma_{ac}\leq$	Comments
1	4.42	C	$\alpha+\beta j\omega$ $\alpha>0,\beta>0$	$\alpha/\beta\delta_s$	—	Geometric interpretation similar to Criterion 3.51. For β arbitrarily small, the result is equivalent to the circle theorem.
2	4.43	C_m	$Z_{10}(j\omega)$ Cor.4.332	γ_{10}	—	Geometric interpretation similar to Criterion 3.52.
3	4.44	C_m	$Z^a_{nl}(j\omega)$ Cor.4.351	—	γ_{a1}	Geometric interpretation not found in literature. Can it be the counterpart of Criterion 3.52?
4	4.45	C_m	$Z_2(j\omega) = Z_{10}(j\omega) + Z^a_{nl}(j\omega)$ Cor.4.352	γ_{20}	γ_{a2}	Geometric interpretation not known.
5	4.46	linear	$Z_2(j\omega)$ Cor.4.353	γ_{30}	γ_{a3} γ_{a1}	,,
6	4.47	linear	$Z_{\ell g}(j\omega)$ Lemma 4.37	$\gamma_{c\ell}$	$\gamma_{ac\ell}$	Geometric interpretation not known. When applied to the the constant coefficient system (2.16), the result

No.	Theorem	Class of $\varphi(\cdot), k(t)$	Multiplier function	$0<\gamma_c<$	$0<\gamma_{ac}<$	Comments
						reduces to the Nyquist criterion thus proving the converse of Lemma 3.22.
7	4.48	C_{mo}	$Z(j\omega)$ of Theorem 4.41 without hypothesis (i) of Lemma 4.36.	γ_c	γ_{ac}	A quasi-geometric interpretation of a special version of the theorem is possible. For comments on the work of Freedman [F6(a)] see Sec3.4(b).
8	4.49	C_{mo}, periodic $k(t)$ with period T_o	$Z(j\omega)=1+\beta j\omega + \sum_n (z_n \cos n\omega T_o + j z'_n \sin n\omega T_o)$ with $\sum_i \lvert z_i \rvert + \lvert z'_i \rvert < 1/2$ and β arbitrarily small.	—	—	A geometric interpretation is possible.
9	4.50	linear, $k(t)$ periodic with period T_o.	$Z(j\omega)$ of The. 4.49 with no restriction on $\sum_i \lvert z_i \rvert + \lvert z'_i \rvert$ except that this be finite.	—	—	Geometric interpretation, simpler than that of Willems [W 3(a)].

Remark 4.42. Hypothesis (iii) of Theorem 4.41 implies, as it is, simultaneous upper and lower bounds on $\Theta(t)$ (the normalized rate of variation of $k(t)$), which are local in character. These bounds are flexible in the sense that one can be traded for the other. In Chapter 3, we obtained a global upper (or, based on Exercise 3.6, lower) bound on $\Theta(t)$ in the Lyapunov-Corduneanu framework. Naturally, we would ask : How do we transform the upper and lower local

bounds on $\theta(t)$ of Theorem 4.41 to global bounds ? The question is resolved by making some changes in the proof of Theorem 4.41. Similar changes can be incorporated in the proofs of the theorem listed in Table 4.2.

1. Replace $k(t)$ in the integrand of (4.45) by $h(t)k(t)$ where $h(t) \in \mathcal{K}$, the class of functions defined in Sec.3.4 (Definition 3.41). Let $\rho_1'(T)$ be the new functional.

Suppose we require, in the place of hypothesis (i) of Theorem 4.41, that

(i)$'$ Re $Z(j\omega-\varepsilon)G(j\omega-\varepsilon) \geq 0$ for some (arbitrarily small) $\varepsilon > 0$ and for all real ω . Since $h(t)\exp(-\varepsilon t/2)$ is nonincreasing, invoking Lemma 4.31, we note that hypothesis (i)$'$ guarantees nonnegativity of $\rho_1'(T)$.

2. In order to establish a lower bound on $\rho_1'(T)$, we need to use Lemma 4.36 with $k(t)$ in the integrand of (4.35) replaced by $h(t)k(t)$. Hence hypothesis (iii) of Theorem 4.41 gives way to

(iii)$'$ $h(t)k(t)\exp(-\gamma_c' t)$ is nonincreasing and $h(t)k(t)\exp(\gamma_{ac} t)$ nondecreasing.

Here $\gamma_c' = \min((1-\alpha)/\beta\delta_s, \gamma_c)$ and γ_{ac} are evaluated from hypothesis (ii) of Theorem 4.41..

Motivated by the proof of Lemma 3.42, we establish the following result, which may be of independent interest as it happens to be a considerable generalization of the 'time varying gain factorization' lemma of Freedman and Zames. [F 6] .

Lemma 4.43. If there exists a time multiplier function' $h(\cdot) \in \mathcal{K}$ such that for some nonnegative constants γ_c' and γ_{ac}, $h(t)k(t)\exp(-\gamma_c' t)$ is nonincreasing and $h(t)k(t)\exp(\gamma_{ac} t)$ is nondecreasing for all $t \geq 0$, then for some positive constants N_1 and N_2 and for all finite $T > 0$ and all $t_o \geq 0$,

$$(1/T)\int_{t_o}^{t_o+T} \theta^+(t)dt \leq N_1 \ ; \qquad -N_2 \leq (1/T)\int_{t_o}^{t_o+T} \theta^-(t)dt$$

and in addition one of the following two sets of inequalities is satisfied.

Set 1. (i) $\lim_{T\to\infty} (1/T)\int_{t_o}^{t_o+T} \theta^+(t)dt \leq \gamma_c'$;

and (ii) $\lim_{T\to\infty} (1/T) \int_{t_o}^{t_o+T} \theta^-(t)dt \geq -\gamma_{ac}$.

Set 2. (i) for $\gamma_c' < \gamma_{ac}$

$$\lim_{T\to\infty} (1/T) \int_{t_o}^{t_o+T} \theta^-(t)dt \geq \gamma_c' - \gamma_{ac},$$

$$\lim_{T\to\infty} (1/T) \int_{t_o}^{t_o+T} \theta^+(t)dt = 0.$$

(ii) for $\gamma_c' > \gamma_{ac}$,

$$\lim_{T\to\infty} (1/T) \int_{t_o}^{t_o+T} \theta^+(t)dt \leq \gamma_c' - \gamma_{ac},$$

$$\lim_{T\to\infty} (1/T) \int_{t_o}^{t_o+T} \theta^-(t)dt = 0.$$

and (iii) for $\gamma_c' = \gamma_{ac}$, $\theta(t)$ is unrestricted.

Proof. See Appendix 4.5.

Remark 4.43. As mentioned earlier, concerning Theorem 4.41, there is some flexibility in the choice of γ_c' and γ_{ac} in meeting the requirement of time-domain inequality in hypothesis (ii). By virtue of Lemma 4.43, if we were to pick $\gamma_c' = \gamma_{ac}$, then one of the possible requirements (Set 2(iii) in Lemma 4.43) on k(t) implies no bound on $\theta(t)$. For the choice of $\gamma_c' = \gamma_{ac}$, Theorem 4.41 along with the Set 2(iii) bound of Lemma 4.43 (Corollary 4.511 to Criterion 4.51 below) is believed to be a generalization of the circle criterion for $\varphi(\cdot)$ monotone. This also applies to Theorem 4.47 (Table 4.2) with $\gamma_{c\ell} = \gamma_{ac\ell}$.

Remark 4.44. All the stability results are applicable to the finite feedback gain problem after a suitable transformation (Exercise 2.2).

4.5. Geometric Interpretation.

We gather from Table 2 that geometric versions of all the stability results are not available. The simplest geometric interpretation is an extension of the Popov criterion applicable to the case of $\varphi(\cdot) \in C$ (Theorem 4.42). This has already been done in Chapter 3 (Criterion 3.51).

We now attempt a geometric statement of Theorem 4.41 and Theorem 4.49. The procedure for the former is motivated by the work of Freedman and Zames [F 6] and leads to a result more general than theirs; the procedure for the latter leads, in the case of Theorem 4.50, to a geometric criterion simpler than that of Willems [W 3(a)] .

4.51. Construction of a multiplier function for Theorem 4.41.

The main problem is, as explained in Sec.4.2 based on Lemma 3.22 and Remark 3.24, to find a complex function of frequency $M(j\omega)$ such that

(i) the $\arg\{M(j\omega)\}$ lies in the band

$$\begin{aligned} \psi_1(\omega) &= [(-\pi/2)+\varepsilon, \psi(\omega)-\varepsilon] \text{ for } \arg\{G(j\omega)\} \geq 0 \\ \psi_2(\omega) &= [\psi(\omega)+\varepsilon, (\pi/2)-\varepsilon] \text{ for } \arg\{G(j\omega)\} < 0 \end{aligned} \tag{3.5}$$

for some constant $\varepsilon > 0$; and

(ii) the multiplier function $M(j\omega)$ expressed in the form

$$M(j\omega) = 1+\beta j\omega + \sum_i z_{ci}\exp(-j\omega\tau_i)+ \sum_{i'} z_{aci'}\exp(j\omega\tau'_{i'})+Z_c(j\omega)+Z_{ac}(j\omega)$$

where $z_{ci}, z_{aci}, Z_c(j\omega), Z_{ac}(j\omega)$ are as defined in Lemma 4.36, satisfies hypothesis (ii) of Theorem 4.41, which in turn has been stated as equivalent to hypotheses (i) and (ii) of Lemma 4.36.

We now examine two extreme cases.

Case 1. It is assumed that there is a minimum finite frequency W beyond which $|\arg\{G(j\omega)\}| < (\pi/2)$ for all $|\omega| > W$.

Case 2. It is assumed that $\arg\{G(j\omega)\}$ tends to $-\pi$ as $\omega \to \infty$.

The other extreme case of $\arg G(j\omega)$ tending to $+\pi$ as $\omega \to \infty$ can be subsumed under Case 2 (Exercise 4.8).

(a) Construction of $M(j\omega)$ for Case 1. Let $U_o(\omega)$ be a real valued continuous almost everywhere differentiable function chosen in the band (3.5) for $|\omega| < W$, and outside the interval $(-W,W)$, let $U_o(\omega) = 0$. Setting $U_o(\omega) = -U_o(-\omega)$, it becomes an odd function of ω. It can be concluded that $U_o(\omega)$ and $dU_o/d\omega$ are in $L_2(-\infty,\infty)$ and hence the inverse (limit-in-the-mean) Fourier transform $u_o(t)$ of $jU_o(\omega)$ is real, odd and in $L_1(-\infty,\infty)$ [T 6, Theorem 68, p 92]. Define $u_e(t)$ on $(-\infty,\infty)$ as follows :

$$\begin{aligned} u_e(t) &= \mu u_o(t) \text{ for } t \geq 0 \\ u_e(t) &= -\mu u_o(t) \text{ for } t < 0 \end{aligned} \tag{4.51}$$

for some constant μ. Then $u_e(t)$ is real, even and is in $L_1(-\infty,\infty)$. Its Fourier transform $U_e(\omega)$ is even. Now let

$$\begin{aligned} \lambda(t) = u_e(t) + u_o(t) &= (\mu+1)u_o(t) \text{ for } t \geq 0, \\ &= (-\mu+1)u_o(t) \text{ for } t < 0 \end{aligned} \tag{4.52}$$

Its Fourier transform is given by

$$\Lambda(j\omega) = U_e(\omega) + j\, U_o(\omega) \tag{4.53}$$

Further, let $(\lambda * \lambda)(t)$ denote the convolution product :

$$(\lambda * \lambda)(t) = \int_{-\infty}^{\infty} \lambda(\tau)\lambda(t-\tau)d\tau$$

and let the subscript 1 in a norm denote the L_1-norm as for instance in

$$\|\lambda\|_1 = \int_{-\infty}^{\infty} |\lambda(t)|dt .$$

We now generate a multiplier function as follows :

$$\begin{aligned} M(j\omega) &= 1 + \Lambda(j\omega) + (\Lambda^2(j\omega)/2!) + \cdots + (\Lambda^n(j\omega)/n!) + \cdots \\ &= \exp(\Lambda(j\omega)) \end{aligned} \tag{4.54}$$

For details on convergence see Freedman and Zames [F 6, Lemma 2, pp 496-497]. It is easy to verify that $\arg M(j\omega) = U_o(\omega)$ and hence in view of the construction of $U_o(\omega)$ so as to lie in the band (3.5), hypothesis (i) of Theorem 4.41 is satisfied. It remains to verify hypothesis (ii) of the theorem. To this end, let

$$\begin{aligned} m_1(t) &= \text{Inverse Fourier transform of } \exp(\Lambda(j\omega))-1 \text{ for } t \geq 0, \\ m_2(t) &= \text{Inverse Fourier transform of } \exp(\Lambda(j\omega))-1 \text{ for } t < 0, \end{aligned} \tag{4.55}$$

and with μ chosen suitably (see below) in (4.51), evaluate γ_c and $\gamma_{ac} > 0$ such that

$$\| m_1 \exp(\gamma_c t)\|_1 + \| m_2 \exp(-\gamma_{ac} t)\|_1 \leq 1/(1+\delta_s-\delta_i). \tag{4.56}$$

Note that an addition of $\beta j\omega$ to (4.54) for arbitrarily small positive β does not affect verification of hypothesis (i) of Theorem 4.41 and hence γ_c' in hypothesis (iii) can be assumed to be equal to γ_c.

The following criterion is a quasi-geometric version of Theorem 4.41.

Criterion 4.51. If $|\arg G(j\omega)| < (\pi/2)$ for all $|\omega| > W$, a finite number, and there exists a function $U_o(\omega)$ in the band (3.5) within the interval $(-W,W)$ and equal to zero outside this interval, such that, with $u_o(t)$ = inverse Fourier transform of $j\,U_o(\omega)$,

(i) $m_1(t)$ and $m_2(t)$ defined by (4.55) and (4.54) are nonpositive ;

(ii) inequality (4.56) holds for some positive γ_c and γ_{ac} ; and

(iii) hypothesis (iii) of Theorem 4.41 (equivalently, the generalized version, Set 1 or Set 2 bound of Lemma 4.43) is satisfied for γ_c' replaced by γ_c, then the system (2.6) for $\varphi(\cdot) \in C_m$ has solutions with Property (4.44)

Since Theorem 4.41 holds for $\varphi(\cdot) \in C_{mo}$ (Table 4.2, Theorem 4.48) by removing the nonpositivity restriction on z_c, z_{ac}, $z_c(\cdot)$ and $z_{ac}(\cdot)$, a simpler geometric interpretation (for Case 1) can be given.

Corollary 4.511. With $U_o(\omega)$ chosen in the band (3.5), $(\tan U_o(\omega))$ is finite in $(-W,W)$, and zero outside this interval. Define $M(j\omega) = 1 + j \tan U_o(\omega) + \beta\, j\omega$ for some arbitrarily small positive β. We note that $(\tan U_o(\omega))$ and its derivative are in $L_2(-\infty,\infty)$ and hence the inverse (limit-in-the-mean) Fourier transform $m_o(t)$ of $(j \tan U_o(\omega))$ is real, odd and is in $L_1(-\infty,\infty)$. As defined by (4.55), $m_1(t) = m_o(t)$ and $m_2(t) = -m_o(-t)$. If there exists a positive constant γ such that

$$\int_0^\infty |m_o(t)| \exp(\gamma t)dt \leq 1/2(1+\delta_s-\delta_i)$$

and for this value of γ, hypothesis (iii) of Theorem 4.41 (equivalently, the generalized version, Set 1 or Set 2 of Lemma 4.43) is satisfied with γ_c and γ_{ac} replaced by γ, then the system (2.6) for $\varphi(\cdot) \in C_{mo}$ has solutions with property (4.44).

Remark 4.51. For $\mu = 1$, the multiplier function $M(j\omega)$ is the one used by

Freedman and Zames [F 6] in which some additional constraints are imposed. The multiplier of [F 6] is a causal function, i.e., $m_2(t) = 0$ for $t < 0$. In this case $\gamma_{ac} = \infty$, implying thereby that the behaviour of the negative lobes of $\theta(t)$ is unrestricted.

In a similar manner, for $\mu = -1$, $m_1(t) = 0$ for $t \geq 0$ and $\gamma_c = \infty$. This means that the behaviour of the positive lobes of $\theta(t)$ is unrestricted. Consequently, for those extreme cases, while using Lemma 4.43 in place of hypothesis (iii) of Theorem 4.41, only the Set 1 bounds on $\theta(t)$ are meaningful. For values of μ in $(-1, 1)$, from inequality (4.56) there is obviously a trade-off between γ_c and γ_{ac}: the larger the value of γ_c, the smaller is the value of γ_{ac} and vice versa. Note also that no practical advantage is gained by choosing $|\mu| > 1$.

(b) Construction of $M(j\omega)$ for Case 2. Find the minimum value of the constant β such that $\arg((1+j\beta\omega)G(j\omega))$ lies in the band $[0,\pi/2)$ for $|\omega| > W$, a finite number. Let $M(j\omega) = (\exp(\Lambda(j\omega)) + \beta j\omega)$ with $\Lambda(j\omega)$ constructed as for Case 1 but with $G(j\omega)$ replaced by $(1+j\beta\omega)G(j\omega)$ for $|\omega| < W$. Because of the additional term in the multiplier, Criterion 4.51 needs a minor change which is suggested as an exercise (Exercise 4.9).

4.52. Construction of a multiplier function for Theorems 4.49 and 4.50.

As before, we consider Case 1 concerning the behaviour of $\arg G(j\omega)$, and leave Case 2 as an exercise. In view of the fact that the multilpier function $Z(j\omega) = 1+\beta j\omega + \sum_n (z_n \cos n\omega T_o + j z'_n \sin n\omega T_o)$ for some arbitrarily small positive β has been used in establishing Theorem 4.49 ($\varphi(\cdot) \in C_{mo}$), Corollary 4.511 above can be straightaway applied. Here we assume that $z_n = 0$ for all $n = 1,2,3,\cdots$.

Since $\tan U_o(\omega)$ is an odd function of ω, with $U_o(\omega)$ chosen in the band (3.5), it remains to derive the Fourier time expansion $\sum_{n=1} z'_n \sin n\omega T_o$ for $\tan U_o(\omega)$ and to evaluate $\sum_n |z'_n|$. If there exists one such expansion for which $\sum_n |z'_n| < (1/2)$, then the system (2.6) with periodic $k(t)$ of period T_o, and $\varphi(\cdot) \in C_{mo}$ has solutions with property (4.44). See Exercise 4.10. For the linear system (2.22) with periodic k(t) of period T_o, Theorem 4.50 as interpreted geometrically in the above manner implies no restriction on $\sum_n |z'_n|$ except that this be finite.

4.53. Comparison with other results in the literature.

Table 4.2 contains some comments on the literature vis-à-vis the results of the present chapter. The form of the constraint ('logarithmic variation') on $\theta(t)$ (derived for the L_2-stability of the system (2.6)) found in [S 14(a)(b)] is reproduced from Freedman and Zames [F 6]. This form of the constraint has already been compared with the present 'global variation' form in Chapter 3 (Sec. 3.4(b)). Further, in [S 14(a)(b)] only the local upper and lower bounds on $\theta(t)$ are interchangeable. In fact, a footnote in [S 14(a)] conjectures the nonexistence (for the system (2.6)) of interchangeable global upper and lower bounds on $\theta(t)$ of the type established in the present chapter.

4.6. Conclusions.

What emerges from an application of the Popov method to the stability analysis of nonlinear and linear time varying systems described by (2.6) and (2.22) respectively is that more general exponential stability conditions can be derived. These conditions include, as a special case, the results of Chapter 3.

As regards the application of the criteria of the chapter to actual problems, much remains to be done. (i) Exercise 3.18, when solved using the criteria of Chapter 3, leads to a bound of the type $\theta(t) \leq 0$ or $\theta(t) \geq 0$. Can one, for this problem, establish exponential stability bounds of the type $-\eta_1 \leq \theta(t) \leq \eta_2$ or of the type given in Lemma 4.43 based on the results of the present chapter?

(ii) What is the complete geometric interpretation of Theorem 4.41 in terms of merely the phase angle characteristic of $G(j\omega)$?

(iii) Do we get improved stability bounds on applying Theorem 4.50 to the (lightly damped) Mathieu equation ?

(iv) Theorem 4.41 generalizes the result of O'Shea [O 2] to the nonlinear time varying system (2.6). The anticausal part of the multiplier in the theorem has resulted in a lower bound on $\theta(t)$. This is the significance of the multiplier functions used in the chapter. Suppose, we specialize the results to hold for (linear and nonlinear) time invariant systems. We can then ask : Is it true that a causal multiplier function $Z_{gc}(j\omega) = 1+\beta\, j\omega + \sum_i z_{ci} \exp(-ij\omega\tau_i) + Z_c(j\omega)$, where z_{ci} and $Z_c(\cdot)$ are as defined in Lemma 4.36, and $\beta > 0$, satisfying the real part conditions (Re $Z_{gc}(j\omega) \geq 0$ and Re $Z_{gc}(j\omega)G(j\omega) \geq 0$ for all real ω), and obeying the time domain constraint $\sum_i |z_{ci}| + \int_0^\infty |z_c(t)|\, dt < 1$ cannot be constructed whereas a noncausal multiplier function of the type used in Lemma 4.36 can be constructed satisfying the real part conditions and the corresponding time domain constraint ? In other words, is a noncausal multiplier function essential in improving the stability results of Chapter 3 as applied to the nonlinear time invariant system (2.24) ?

The next chapter contains a description of the Zames - Sandberg method for the L_2-stability analysis of the system represented by (2.7) under Assumption S2 (Sec.2.32). The stability results so derived constitute some generalization of the results of the present chapter.

Appendix 4.1

Proof of Lemma 4.33.

$$I_{ni} = \int_0^T k(t)\varphi_T(\sigma(t))\sigma_T(t)dt +$$

$$(\nu_i-1)\mu_i \int_0^T k(t)\varphi_T(\sigma(t)) \left(\int_0^t \exp(-\mu_i(t-\tau))\sigma_T(\tau)d\tau\right)dt \qquad (4.12)$$

Because $\varphi(\cdot)\in C_m$, we have

$$(\varphi(\sigma_1)-\varphi(\sigma_2))(\sigma_1-\sigma_2) \geq 0 \text{ for all } \sigma_1 \text{ and } \sigma_2. \qquad (4.57)$$

In (4.12), let $\sigma_1 = \sigma_T(t)$ and

$$\sigma_1-\sigma_2 = \sigma_T(t) + (\nu_i-1)\ \mu_i \int_0^T \exp(-\mu_i(t-\tau))\sigma_T(\tau)d\tau$$

from which

$$\sigma_2 = -\mu_i(\nu_i-1) \int_0^t \exp(-\mu_i(t-\tau))\sigma_T(\tau)d\tau$$

and $$(d\sigma_2/dt) = -\mu_i\sigma_2 + \mu_i(1-\nu_i)\sigma_T(t) \tag{4.58}$$

or $$\sigma_1 = ((d\sigma_2/dt) + \mu_i\sigma_2)/\mu_i(1-\nu_i)$$

Adding to and subtracting from (4.12) the expression $\varphi_T(\sigma_2)(\sigma_1-\sigma_2)$, we have

$$I_{ni} = \int_0^T k(t)(\varphi_T(\sigma_1)-\varphi_T(\sigma_2))(\sigma_1-\sigma_2)dt + \int_0^T k(t)\varphi_T(\sigma_2)(\sigma_1-\sigma_2)dt \tag{4.59}$$

The first integral of (4.59) is nonnegative by virtue of (4.57) ; the second integral on using (4.58) becomes

$$\int_0^T k(t)\varphi_T(\sigma_2)(\sigma_1-\sigma_2)dt = (1/(1-\nu_i)\mu_i)\int_0^T k(t)\varphi_T(\sigma_2)((d\sigma_2/dt)+\nu_i\mu_i\sigma_2)dt \tag{4.60}$$

Invoke Lemma 4.32 to conclude that the integral (4.60) is nonnegative if $k(t)\exp(-\gamma_{1i}t)$ is nonincreasing for some nonnegative $\gamma_{1i} \leq (\nu_i\mu_i/\delta_s)$. The lemma is proved.

Appendix 4.2

Proof of Lemma 4.36.

$$I_{ng} = \int_0^T k(t)\varphi_T(\sigma(t))\Big\{\alpha\sigma_T(t) + \sum_i z_{ci}\sigma_T(t-\tau_i) + \sum_{i'} z_{aci'}\sigma_T(t+\tau'_{i'})$$
$$+ \int_0^\infty z_c(\tau)\sigma_T(t-\tau)d\tau + \int_{-\infty}^0 z_{ac}(\tau)\sigma_T(t-\tau)d\tau\Big\}dt$$

Suppose $k(t)\exp(-\gamma_c t)$ is nonincreasing and $k(t)\exp(\gamma_{ac}t)$ is nondecreasing for some nonnegative constants γ_c and γ_{ac}. Then from Lemma 4.31, there are T_1 and T_2 such that $0 \leq T_1, T_2 \leq T$ and

$$I_{ng} = k(0)\int_0^{T_1} \exp(\gamma_c t)\varphi_T(\sigma(t))\Big\{c\sigma_T(t) + \sum_i z_{ci}\sigma_T(t-\tau_i) + \int_0^\infty z_c(\tau)\sigma_T(t-\tau)d\tau\Big\}dt$$
$$+ k(T)\exp(\gamma_{ac}T)\int_{T_2}^T \exp(-\gamma_{ac}t)\varphi_T(\sigma(t))\Big\{(1-c)\sigma_T(t) + \sum_{i'} z_{aci'}\sigma_T(t+\tau'_{i'})$$
$$+ \int_{-\infty}^0 z_{ac}(\tau)\sigma_T(t-\tau)d\tau\Big\}dt \tag{4.61}$$

for some c satisfying the inequality $0 < c < \alpha$.

In (4.61) we consider the two integrals separately :

$$I_{ng_1} = \int_0^{T_1} \exp(\gamma_c t)\varphi_T(\sigma(t)) \left\{ c\sigma_T(t) + \sum_i z_{ci}\sigma_T(t-\tau_i) + \int_0^{\infty} z_c(\tau)\sigma_T(t-\tau)d\tau \right\} dt. \quad (4.62)$$

Assuming that the order of performing the integration in the last integral of (4.62) can be interchanged, we get

$$I_{ng_1} = c\int_0^{T_1} \exp(\gamma_c t)\varphi_T(\sigma(t))\sigma_T(t)dt + \sum_i z_{ci}\exp(\gamma_c\tau_i)\int_0^{T_1}\exp(\gamma_c(t-\tau_i))\varphi_T(\sigma(t))$$

$$\sigma_T(t-\tau_i)dt + \int_0^{\infty} z_c(\tau)\exp(\gamma_c\tau)\left\{\int_0^{T_1}\varphi_T(\sigma(t))\sigma_T(t-\tau)\exp(\gamma_c(t-\tau))dt\right\} d\tau \quad (4.63)$$

We consider the last two integrals of (4.63) separately in an attempt to find their lower bounds. These lower bounds referred to the first integral of (4.63) will then give us the condition(s) for the nonnegativity of I_{ngl}.

Let
$$J_{nl} = \int_0^{T_1} \exp(\gamma_c(t-\tau_i))\varphi_T(\sigma(t))\sigma_T(t-\tau_i)dt \quad (4.64)$$

From the monotonicity property of $\varphi(\cdot)$, we have

$$\varphi(v_1)(v_1-v_2) \geqslant \Phi(v_1) - \Phi(v_2) \text{ for all } v_1 \text{ and } v_2 \quad (4.65)$$

With respect to (4.64), define $v_1 = \sigma_T(t)$ and $v_2 = \sigma_T(t-\tau_i)\exp(-\gamma_c\tau_i)$.
Then, noting that $\varphi_T(\sigma(t)) = \varphi_T(\sigma_T(t))$,

$$\int_0^{T_1}\exp(\gamma_c t)\varphi_T(\sigma(t))(\sigma_T(t)-\sigma_T(t-\tau_i)\exp(-\gamma_c\tau_i))dt \geqslant$$

$$\int_0^{T_1} \exp(\gamma_c t)\Phi(\sigma_T(t))dt - \int_0^{T_1} \exp(\gamma_c t)\Phi(\sigma_T(t-\tau_i)\exp(-\gamma_c\tau_i))dt \quad (4.66)$$

The last integral of (4.66) can be rewritten as follows by changing the variable integration :

$$\int_0^{T_1}\exp(\gamma_c t)\Phi(\sigma_T(t-\tau_i)\exp(-\gamma_c\tau_i))dt = \int_{-\tau_i}^{T_1-\tau_i} \exp(\gamma_c(t+\tau_i))\Phi(\sigma_T(t)\exp(-\gamma_c\tau_i))dt$$

$$\leqslant \int_0^{T_1}\exp(\gamma_c(t+\tau_i))\Phi(\sigma_T(t)\exp(-\gamma_c\tau_i))dt$$

$$\leqslant \delta_s \int_0^{T_1} \exp(\gamma_c t)\varphi_T(\sigma(t))\sigma_T(t)dt \quad (4.67)$$

Further $\Phi(\sigma_T(t)) \geqslant \delta_i\, \varphi_T(\sigma(t))\sigma_T(t).$ (4.68)

Hence using the inequalities (4.67) and (4.68) in (4.66), we get

$$\int_0^{T_1} \exp(\gamma_c(t-\tau_i))\varphi_T(\sigma(t))\sigma_T(t-\tau_i)dt \leq (1+\delta_s-\delta_i) \int_0^{T_1} \exp(\gamma_c t)\varphi_T(\sigma(t))\sigma_T(t)dt \quad (4.69)$$

Next, consider the integral

$$J_{n2} = \int_0^{T_1} \varphi_T(\sigma(t))\sigma_T(t-\tau)\exp(\gamma_c(t-\tau))dt. \quad (4.70)$$

Define $v_1 = \sigma_T(t)$ and $v_2 = \sigma_T(t-\tau)\exp(-\gamma_c\tau)$. Then from the monotonicity property of (4.65) of $\varphi(\cdot)$, we have

$$\int_0^{T_1}\exp(\gamma_c t)\varphi_T(\sigma(t))(\sigma_T(t) - \sigma_T(t-\tau)\exp(-\gamma_c\tau))dt$$

$$\geq \int_0^{T_1} \exp(\gamma_c t)\Phi(\sigma_T(t))dt - \int_0^{T_1} \exp(\gamma_c t)\Phi(\sigma_T(t-\tau)\exp(-\gamma_c\tau))dt \quad (4.71)$$

The last integral of (4.71) can be rewritten as follows by changing the variable of integration :

$$\int_0^{T_1} \exp(\gamma_c t)\Phi(\sigma_T(t-\tau)\exp(-\gamma_c\tau))dt = \int_{-\tau}^{T_1-\tau} \exp(\gamma_c(t+\tau))\Phi(\sigma_T(t)\exp(-\gamma_c\tau))dt$$

$$\leq \int_0^{T_1} \exp(\gamma_c(t+\tau))\Phi(\sigma_T(t)\exp(-\gamma_c\tau))dt$$

$$\leq \delta_s \int_0^{T_1} \exp(\gamma_c t)\varphi_T(\sigma(t)\exp(-\gamma_c\tau))\sigma_T(t)dt$$

$$\leq \delta_s \int_0^{T_1} \exp(\gamma_c t)\varphi_T(\sigma(t))\sigma_T(t)dt \quad (4.72)$$

Further $\int_0^{T_1}\exp(\gamma_c t)\Phi(\sigma_T(t))dt \geq \delta_i \int_0^{T_1} \exp(\gamma_c t)\varphi_T(\sigma(t))\sigma_T(t)dt$ (4.73)

Substituting the inequalities (4.72) and (4.73) in (4.71), we get

$$\int_0^{T_1} \exp(\gamma_c(t-\tau))\varphi_T(\sigma(t))\sigma_T(t-\tau)dt \leq (1+\delta_s-\delta_i) \int_0^{T_1} \exp(\gamma_c t)\varphi_T(\sigma(t))\sigma_T(t)dt \quad (4.74)$$

The inequalities (4.69) and (4.74) when applied to (4.63) give us the following basic result:

I_{ng1} is nonnegative for all $\sigma_T(t)$ if

(i) $z_{ci} \leq 0$ for all i and $z_c(t) \leq 0$ for all $t \geq 0$; and

$$\text{(ii)} \sum_i |z_{ci}| \exp(\gamma_c \tau_i) + \int_0^{\infty} |z_c(\tau)| \exp(\gamma_c \tau) d\tau \leq c/(1+\delta_s-\delta_i). \qquad (4.75)$$

If $\varphi(\cdot) \in C_{mo}$, then condition (i) can be removed.

We now consider the second integral of (4.61). Let

$$I_{ng2} = \int_{T_2}^{T} \exp(-\gamma_{ac} t) \varphi_T(\sigma(t)) \Big\{ (\alpha-c)\sigma_T(t) + \sum_{i'} z_{aci'} \sigma_T(t+\tau'_{i'}) + \int_{-\infty}^{0} z_{ac}(\tau)\sigma_T(t-\tau) d\tau \Big\} dt \qquad (4.76)$$

Assuming that the order of performing the integration in the last integral of (4.76) can be interchanged, we get

$$I_{ng2} = (\alpha-c) \int_{T_2}^{T} \exp(-\gamma_{ac} t) \varphi_T(\sigma(t)) \sigma_T(t) dt$$

$$+ \sum_{i'} z_{aci'} \exp(\gamma_{ac} \tau'_{i'}) \int_{T_2}^{T} \exp(-\gamma_{ac}(t+\tau'_{i'})) \varphi_T(\sigma(t)) \sigma_T(t+\tau'_{i'}) dt$$

$$+ \int_{-\infty}^{0} z_{ac}(\tau) \exp(-\gamma_{ac}\tau) \Big\{ \int_{T_2}^{T} \exp(-\gamma_{ac}(t-\tau)) \varphi_T(\sigma(t)) \sigma_T(t-\tau) dt \Big\} d\tau \qquad (4.77)$$

We consider, as before, the last two integrals of (4.77) separately in an attempt to find their lower bounds. These lower bounds referred to the first integral of (4.77) will then give us the conditions for the nonnegativity of I_{ng2}.

Let
$$J_{n3} = \int_{T_2}^{T} \exp(-\gamma_{ac}(t+\tau'_{i'})) \varphi_T(\sigma(t)) \sigma_T(t+\tau'_{i'}) dt \qquad (4.78)$$

From the monotonicity of $\varphi(\cdot)$, using inequality (4.65), we get

$$\int_{T_2}^{T} \exp(-\gamma_{ac} t) \varphi_T(\sigma(t)) (\sigma_T(t) - \sigma_T(t+\tau'_{i'}) \exp(-\gamma_{ac}\tau'_{i'})) dt$$

$$\geq \int_{T_2}^{T} \exp(-\gamma_{ac} t) \Phi(\sigma_T(t)) dt - \int_{T_2}^{T} \exp(-\gamma_{ac} t) \Phi(\sigma_T(t+\tau'_{i'}) \exp(-\gamma_{ac}\tau'_{i'})) dt \qquad (4.79)$$

By changing the variable of integration, the last integral of (4.79) can be rewritten as

$$\int_{T_2}^{T} \exp(-\gamma_{ac}t)\Phi(\sigma_T(t+\tau'_{i'})\exp(-\gamma_{ac}\tau'_{i'}))dt =$$

$$\int_{T_2+\tau'_{i'}}^{T+\tau'_{i'}} \exp(-\gamma_{ac}(t-\tau'_{i'}))\Phi(\sigma_T(t)\exp(-\gamma_{ac}\tau'_{i'}))dt \tag{4.80}$$

Noting that $\sigma_T(t) = 0$ for $t > T$, inequality (4.80) reduces to

$$\int_{T_2}^{T_1} \exp(-\gamma_{ac}t)\Phi(\sigma_T(t+\tau'_{i'})\exp(-\gamma_{ac}\tau'_{i'}))dt \leq \delta_s \int_{T_2}^{T} \exp(-\gamma_{ac}t)\varphi_T(\sigma(t))\sigma_T(t)dt \tag{4.81}$$

Hence, from (4.79), we conclude that

$$\int_{T_2}^{T} \exp(-\gamma_{ac}(t+\tau'_{i'}))\varphi_T(\sigma(t))\sigma_T(t+\tau'_{i'})dt \leq (1+\delta_s-\delta_i)\int_{T_2}^{T} \exp(-\gamma_{ac}t)\varphi_T(\sigma(t))\sigma_T(t)dt \tag{4.82}$$

Proceeding along the same lines for the last integral of (4.77), we find that

$$\int_{T_2}^{T} \exp(-\gamma_{ac}(t-\tau))\varphi_T(\sigma(t))\sigma_T(t-\tau)dt \leq (1+\delta_s-\delta_i)\int_{T_2}^{T} \exp(-\gamma_{ac}t)\varphi_T(\sigma(t))\sigma_T(t)dt \tag{4.83}$$

Use of inequalities (4.82) and (4.83) in (4.77) leads to the following basic result :

I_{ng2} is nonnegative for all $\sigma_T(t)$ if

(i) $z_{aci} \leq 0$ for all i and $z_{ac}(t) \leq 0$ for all $t \leq 0$; and

(ii) $\sum_i |z_{aci}| \exp(\gamma_{ac}\tau'_i) + \int_{-\infty}^{0} |z_{ac}(\tau)| \exp(-\gamma_{ac}\tau)d\tau \leq (\alpha-c)/(1+\delta_s-\delta_i)$.

If $\varphi(\cdot) \in C_{mo}$, then condition (i) can be removed.

By virtue of our definitions, $I_{ng} = k(0)I_{ng1} + k(T)\exp(\gamma_{ac}T)I_{ng2}$.

Therefore I_{ng} is nonnegative for all $\sigma_T(t)$ if the hypotheses of the lemma are satisfied. The lemma is proved.

Corollary 4.361. If $\varphi(\cdot) \in C_{mo}$, then Lemma 4.36 holds without hypothesis (i).

Appendix 4.3

Proof of Lemma 4.38.

$$I_{np} = \int_0^T k(t)\varphi_T(\sigma(t))\sigma_T(t)dt + \sum_i z_{ci} \int_0^T k(t)\varphi_T(\sigma(t))\sigma(t-iT_o)dt$$

$$+ \sum_{i'} z_{aci'} \int_0^T k(t)\varphi_T(\sigma(t))\sigma_T(t+i' T_o)dt \qquad (4.84)$$

In the last integral of (4.84), let $t+i' T_o = \tau$. Then this term becomes

$$J_{pl} = \sum_{i'} z_{aci'} \int_0^T k(t)\varphi_T(\sigma(t))\sigma_T(t+i'T_o)dt =$$

$$\sum_{i'} z_{aci'} \int_{i'T_o}^{T+i'T_o} k(\tau-i'T_o)\varphi_T(\sigma(\tau-i'T_o))\sigma_T(\tau)d\tau$$

Since $\sigma_T(t) = 0$ for $t < 0$ and for $t > T$, and $k(\tau-i'T_o) = k(\tau)$, we have

$$J_{pl} = \sum_{i'} z_{aci'} \int_0^T k(t)\varphi_T(\sigma(t-i'T_o))\sigma_T(t)dt.$$

Therefore

$$I_{np} = \int_0^T k(t)\varphi_T(\sigma(t))\sigma_T(t)dt + \sum_i z_{ci} \int_0^T k(t)\varphi_T(\sigma(t))\sigma(t-iT_o)dt$$

$$+ \sum_{i'} z_{aci'} \int_0^T k(t)\varphi_T(\sigma(t-i'T_o))\sigma_T(t)dt \qquad (4.85)$$

We now consider the case $z_{ci} = z_{aci}$ for all $i = 1,2,3,\cdots,$. From the monotonicity property of $\varphi(\cdot)$,

$$\varphi_T(\sigma(t))\sigma_T(t-iT_0)+\varphi_T(\sigma(t-iT_o))\sigma(t) \leq \varphi_T(\sigma(t))\sigma_T(t)+\varphi_T(\sigma(t-iT_o))\sigma_T(t-iT_o)$$

Hence, in view of the periodic nature of $k(t)$,

$$\int_0^T k(t)\left\{\varphi_T(\sigma(t))\sigma(t-T_o)+\varphi_T(\sigma(t-iT_o))\sigma_T(t)\right\}dt \leq 2\int_0^T k(t)\varphi_T(\sigma(t))\sigma_T(t)dt$$

Therefore from (4.85), I_{np} is nonnegative for $z_{ci} = z_{aci}$ for all $i = 1,2,3,$ $\cdots$, if (i) $z_{ci} \leq 0$ for all $i = 1,2,3,\cdots$, and (ii) $\sum_i |z_{ci}| \leq 1/2$.

For $z_{ci} \neq z_{aci}$ for all $i = 1,2,3,\cdots$, a result of the above type cannot be established when $\varphi(\cdot) \in C_m$. However, when $\varphi(\cdot) \in C_{mo}$, naturally the above result holds and, in addition, it is possible to consider the case

$z_{ci} = -z_{aci}$ for all $i = 1,2,3,\cdots$. In this case one has to use the additional property of $\varphi(\cdot) \in C_{mo}$, namely,

$$\varphi(\sigma_1)(\sigma_1+\sigma_2) - \varphi(\sigma_2)(\sigma_1-\sigma_2) \geq 0$$

for all σ_1 and σ_2. The following result is true.

Let z_i' be a real constant for $i = 1,2,3,\cdots$, such that $\sum_i |z_i'| < \infty$. Define

$$(Z_p' \sigma_T)(t) = \sum_i z_i' \left\{\sigma_T(t-iT_o)-\sigma_T(t+iT_o)\right\} + \sigma_T(t).$$

Then the functional

$$I_{np}' = \int_0^T k(t)\varphi_T(\sigma(t))(Z_p'\sigma_T)(t)dt$$

is nonnegative for $\varphi(\cdot) \in C_{mo}$ and periodic $k(t)$ with period T_o, if

$$\sum_i |z_i'| \leq 1/2 .$$

Appendix 4.4

Evaluation of the upper bound on $\rho_1(T)$ of Theorem 4.41.

$$I_u = \int_0^T \left\{(1-\alpha)f_T(t)+\beta(df_T(t)/dt)+ (Z_g f_T)(t)\right\} k(t)\varphi_T(\sigma(t))dt \tag{4.86}$$

Given that $|f(t)|$ and $|df(t)/dt|$ are bounded by $r_1 \exp(-r_o t)$ for some positive constants r_1 and r_o, and for all $t \geq 0$, the verification of the following inequalities is straightforward.

$$|(1-\alpha)f_T(t) + \beta(df_T(t)/dt)| \leq r_3 \exp(-r_2 t),$$
$$|(Z_g f_T)(t)| \leq r_5 \exp(-r_4 t)$$

for some positive constants r_2-r_5, and for all $t \geq 0$. Hence the modulus of the term inside the curly brackets of (4.86) is bounded by $r_7 \exp(-r_6 t)$ for some positive constants r_6 and r_7 and for all $t \geq 0$.

Consequently,

$$|I_u| \leq \int_0^T r_1(\exp(-r_6 t)k(t)\varphi_T(\sigma(t))dt$$

$$\leq \eta \sup_{0 \leq t \leq T} |\sigma(t)| \int_0^T \exp(-r_6 t)dt$$

$$\leq R_1 \sup_{0\leq t\leq T} |\sigma(t)|$$

for some constant R_1 independent of T. The lemma is proved.

Appendix 4.5

Proof of Lemma 4.43. We have (hypothesis (iii)$'$ of Theorem 4.41)

$$-\gamma_{ac} \leq (\tfrac{dh}{dt}/h) + \theta(t) \leq \gamma_c', \quad t\in [0,\infty) \tag{!}$$

Let

$$(\tfrac{dh}{dt}/h) = h_1(t) + h_2(t) \tag{4.87}$$

Then inequality (!) is satisfied by choosing

$$h_1(t) = \gamma_c' - \theta^+(t) \ ; \quad h_2(t) = -\gamma_{ac} - \theta^-(t) \tag{4.88}$$

where $\theta^+(t)$ and $\theta^-(t)$, to recall, are respectively the positive and negative lobes of $\theta(t)$.

We find that $h(t)$ so constructed satisfies the inequality (!) . Then we have to verify whether $h(\cdot)\in \mathcal{K}$. Substitute (4.88) in(!) and solve for $h(t)$ to get

$$h(t) = h(t_o)\exp\Big(\int_{t_o}^{t} (-\gamma_{ac}+ \gamma_c'- \theta^-(\tau)-\theta^+ (\tau))d\tau\Big) \tag{4.89}$$

If $h(\cdot)\in \mathcal{K}$, then for some positive constants η_1 and η_2 with $\eta_1 < \eta_2$,

$$\eta_1 \leq \exp\Big(\int_{t_o}^{t} (-\gamma_{ac}- \gamma_c'- \theta^-(\tau)- \theta^+(\tau))d\tau\Big) \leq \eta_2 \tag{4.90}$$

for all $t_o \geq 0$ and all $t \geq t_o$.

The inequality (4.90) can be satisfied in any one of the following five ways : For some positive constants M_1, M_2, M_3 and M_4, for all $t_o \geq 0$ and all $t \geq t_o$,

Case 1.
$$-\infty < -M_1 \leq \int_{t_o}^{t} (-\gamma_{ac}+ \gamma_c'- \theta(\tau))d\tau \leq M_2 < \infty \tag{4.91}$$

Case 2.(i)
$$-\infty < -M_1 \leq \int_{t_o}^{t} (\gamma_c'- \theta^-(\tau))d\tau \leq M_2 < \infty \ ; \tag{4.92}$$

and (ii)
$$-\infty < -M_3 \leq \int_{t_o}^{t} (-\gamma_{ac}- \theta^+(\tau))d\tau \leq M_4 < \infty \tag{4.93}$$

Case 3. (i) $$-\infty < -M_1 \leq \int_{t_o}^{t} (-\gamma_{ac} - \theta^-(\tau))d\tau \leq M_2 < \infty ; \tag{4.94}$$

and (ii) $$-\infty < -M_3 \leq \int_{t_o}^{t} (\gamma_c' - \theta^+(\tau))d\tau \leq M_4 < \infty ; \tag{4.95}$$

Case 4. (i) $$-\infty < -M_1 \leq \int_{t_o}^{t} (\gamma_c' - \gamma_{ac} - \theta^-(\tau))d\tau \leq M_2 < \infty ; \tag{4.96}$$

and (ii) $$-\infty < -M_3 \leq \int_{t_o}^{t} (-\theta^+(\tau))d\tau \leq M_4 < \infty \tag{4.97}$$

Case 5. (i) $$-\infty < -M_1 \leq \int_{t_o}^{t} (-\theta^-(\tau))d\tau \leq M_2 < \infty ; \tag{4.98}$$

and (ii) $$-\infty < -M_3 \leq \int_{t_o}^{t} (\gamma_c' - \gamma_{ac} - \theta^+(\tau))d\tau \leq M_4 < \infty \tag{4.99}$$

Case 1. Inequality (4.91) can be reduced to

$$\exp(-M_2 + (\gamma_c' - \gamma_{ac})(t-t_o)) \leq (k(t)/k(t_o)) \leq \exp(M_1 + (\gamma_c' - \gamma_{ac})(t-t_o)) \tag{4.100}$$

for all $t_o \geq 0$ and all $t \geq t_o$.

When $\gamma_c' > \gamma_{ac}$ and when $\gamma_c' < \gamma_{ac}$, we conclude from (4.100) that $k(\cdot) \notin \mathcal{K}$. Hence Case 1 is ruled out for $\gamma_c' \neq \gamma_{ac}$. But when $\gamma_c' = \gamma_{ac}$, inequality (4.100) merely implies that $k(\cdot) \in \mathcal{K}$ and hence no restriction on $\theta(t)$ is imposed.

Case 2. From inequality (4.92),

$$-M_2 + \gamma_c'(t-t_o) \leq \int_{t_o}^{t} \theta^-(\tau)d\tau \leq M_1 + \gamma_c'(t-t_o) \tag{4.101}$$

for all $t_o \geq 0$ and all $t \geq t_o$. But (4.101) is untenable because $\theta^-(\tau)$ is always negative and hence the left hand side of the inequality is violated for sufficiently large $t > t_o$. (The right hand side of (4.101) is, however, trivially satisfied). In an identical manner, inequality (4.93) is untenable. Hence Case 2 is ruled out.

Case 3. From inequalities (4.94) and (4.95) respectively,

$$-M_2 - \gamma_{ac}(t-t_o) \leq \int_{t_o}^{t} \theta^-(\tau)d\tau \leq M_1 - \gamma_{ac}(t-t_o) \tag{4.102}$$

$$-M_4 + \gamma_c'(t-t_o) \leq \int_{t_o}^{t} \theta^+(\tau)d\tau \leq M_3 + \gamma_c'(t-t_o) \tag{4.103}$$

for all $t_o \geq 0$ and all $t \geq t_o$.

Hence by requiring that $\int_{t_o}^{t} \theta^+(\tau)d\tau$ and $\int_{t_o}^{t} |\theta^-(\tau)| d\tau$ be bounded for all finite $t \geq t_o$, and

$$\lim_{T\to\infty} (1/T) \int_{t_o}^{t_o+T} \theta^+(t)dt \leq \gamma_c' ;$$

$$\lim_{T\to\infty} (1/T) \int_{t_o}^{t_o+T} \theta^-(t)dt \geq -\gamma_{ac},$$

inequalities (4.102) and (4.103) are satisfied.

Case 4. From inequalities (4.96) and (4.97) respectively,

$$-M_2+(\gamma_c'-\gamma_{ac})(t-t_o) \leq \int_{t_o}^{t} \theta^-(\tau)d\tau \leq M_1+(\gamma_c'-\gamma_{ac})(t-t_o) \tag{4.104}$$

$$-M_4 \leq \int_{t_o}^{t} \theta^+(\tau)d\tau \leq M_3 \tag{4.105}$$

for all $t_o \geq 0$ and all $t \geq t_o$.

When $\gamma_c' > \gamma_{ac}$, the left hand side of the inequality (4.104) is violated (in view of the negative nature of $\theta^-(\tau)$) for sufficiently large $t > t_o$. Hence inequality (4.104) is incompatible for $\gamma_c' > \gamma_{ac}$. Consequently, we require that $\gamma_c' \leq \gamma_{ac}$. When $\gamma_c' = \gamma_{ac}$, inequality (4.104) gives

$$-M_2 \leq \int_{t_o}^{t} \theta^-(\tau)d\tau \leq M_1 \tag{4.106}$$

for all $t_o \geq 0$ and all $t \geq t_o$. Inequalities (4.105) and (4.106) are definitely more restrictive than the waiving of all restrictions on $\theta(t)$ in Case 1 for $\gamma_c' = \gamma_{ac}$. Hence we need consider only $\gamma_c' < \gamma_{ac}$. As in Case 3, if

$\int_{t_o}^{t} |\theta^-(\tau)| d\tau$ is bounded for all finite $t \geq t_o$, and

$$\lim_{T\to\infty} (1/T) \int_{t_o}^{t} \theta^-(\tau)d\tau \geq \gamma_c' - \gamma_{ac},$$

then (4.105) holds.

Case 5. As in Case 4, inequalities (4.98) and (4.99) are respectively equivalent to requiring that

$$-M_2 \leq \int_{t_o}^{t} \theta^-(\tau)d\tau \leq M_1$$

for all $t_o \geq 0$ and all $t \geq t_o$; and for $\gamma_c' > \gamma_{ac}$,

$$\int_{t_o}^{t} \theta^+(\tau)d\tau \quad \text{is bounded for all } t \geq t_o,$$

$$\lim_{T\to\infty} (1/T) \int_{t_o}^{t_o+T} \theta^+(t)dt \leq \gamma_c' - \gamma_{ac}.$$

When $\gamma_c' < \gamma_{ac}$, $k(\cdot) \notin \mathcal{K}$ and when $\gamma_c' = \gamma_{ac}$, the constraint so obtained on $\theta(t)$ is stronger than necessary as explained in Case 4. The lemma is proved.

EXERCISES

4.1. Find conditions for the passivity of the cascaded connection of $k(t)\varphi(\cdot)$ with $\varphi(\cdot) \in C_{mo}$ and a time invariant linear causal system with the frequency response characteristic

$$Z_{11}(j\omega) = \left\{\alpha + j\beta\omega + \sum_i \eta_i(j\omega + \nu_i\mu_i)/(j\omega + \mu_i)\right\}$$

where $\alpha > 0$, $\beta \geq 0$; $\eta_i \geq 0$; $\mu_i > 0$ and $\nu_i \geq 0$ for all i. Also prove Lemma 4.34.

4.2. Repeat Exercise 4.1 for the cascaded connection of $k(t)\varphi(\cdot)$ with $\varphi(\cdot) \in C_{mo}$ and a time invariant linear noncausal system with the frequency response characteristic

$$Z_{22}(j\omega) = \left\{\alpha + j\beta\omega + \sum_i \eta_i'(\zeta_i \nu_i' - j\omega)/(\zeta_i - j\omega)\right\}$$

where $\alpha > 0$, $\beta \geq 0$; $\eta_i' \geq 0$, $\zeta_i > 0$ and $\nu_i' \geq 0$ for all i .Repeat the exercise for the choice of noncausal frequency response function

$$Z_{23}(j\omega) = \left\{\alpha + j\beta\omega + \sum_i \eta_i(j\omega + \nu_i\mu_i)/(j\omega + \mu_i) + \sum_{i'} \eta_i'(\zeta_i \nu_i' - j\omega)/(\zeta_i - j\omega)\right\}.$$

Also prove Corollary 4.353 (Lemma 4.35).

4.3. Find conditions for the passivity of the cascaded connection of $k(t)\varphi(\cdot)$

with $\varphi(\cdot)\in C_m$, C_{mo} and a time invariant linear (causal+anticausal) system with the frequency response characteristic

$$Z_{31}(j\omega) = \alpha+\beta j\omega + \sum_i \eta_i(\beta_i+j\omega\alpha_i+(j\omega)^2)/(\lambda_i+\mu_i j\omega+(j\omega)^2)$$
$$+ \eta_i'(\beta_i'-j\omega\alpha_i'+(j\omega)^2)/(\lambda_i'-\mu_i' j\omega+(j\omega)^2)$$

4.4. For the class of power law nonlinearities defined in Exercise 3.13, repeat Exercises 4.1-4.3.

4.5. Establish results similar to Lemma 4.36 and Corollary 4.361 for $\varphi(\cdot)\in C_{mo}$.

4.6. Prove Lemma 4.37. Compare the results with those of [S 14(c)].

4.7. Prove Lemma 4.39 on the lines of the proof of Lemma 4.37 or otherwise..

Manley - Rowe type frequency-power formulas are relations between the power inputs and power outputs of a nonlinear device at various frequencies of the almost periodic input. These formulas have found application in the design of frequency convertors Using Lemma 4.39, it is possible to establish results more general than those of Skoog and Willems [S 7] on the frequency-power formula.

4.8. Suppose arg $G(j\omega)$ tends to $+\pi$ as $\omega\to\infty$. What modification do you suggest in the statement of Theorem 4.41 so that the geometric interpretation of Sec.4.51 can be applied to it ?

4.9. Complete the geometric interpretation of Sec 4.51, Case 2.

4.10. Let $G(s) = 40/(s+0.01)(s+1)(s^2+0.8s+16)$. Find the Hurwitz sector of stability for the linear time invariant system (2.15). With the nonlinear time varying feedback gain confined to the Hurwitz sector, $k(t)$ periodic and $\varphi(\cdot)\in C_{mo}$, it is required to determine T_o, the period of $k(t)$, for exponential stability of (2.6). To this end, show that the multiplier function $Z(j\omega) = 1+\beta j\omega + j\,0.999 \sin 1.112\omega$ with $\beta > 0$ arbitrarily small, satisfies the hypotheses of Theorem 4.49 (Table 4.2), and demonstrate that the system (2.6) is exponentially stable for $T_o \leq 1.112$.

4.11. Consider the nonlinear time varying system (2.6) having $G(s) = (s+0.005)^2(s+0.1)^2(s+1000)/(s+0.0001)^2\ (s+2)^2(s+50)^2$. Verify whether

Assumption S1 (Sec.2.32 of Chapter 2) holds. For what values of the constants μ, μ', β, η and η', can the multiplier function

$$Z(j\omega) = 1+\beta j\omega + (\eta/(j\omega+\mu)) + (\eta'/(\mu'-j\omega))$$

be used to satisfy hypothesis (i) of Theorem 4.41 ? Hence find conditions for the exponential stability of (2.6) with $\varphi(\cdot)\in C_m$, C_{mo}.

4.12. For the linear system (2.22) with $G(s) = 1/s(s+2)$ and $k(t)$ periodic with period T_o, show, based on Theorem 4.50, that a trade-off between T_o and the upper bound on $k(t)$ can be established to guarantee exponential stability. Compare with the result of Willems [W 3(a)].

4.13. For the nonlinear time varying system (2.6) with $G(s) = (10s+1)(2s+1)/(s^2+20s+400)(2s^2+5s+4)$ and $\varphi(\cdot)\in C_m$, C_{mo}, it is required to establish exponential stability conditions based on an explicit multiplier function construction outlined in Sec.4.5. Compare the multiplier so derived with $Z(s) = (2s^2+5s+4)/(2s+1)$ as suggested by Brockett and Willems [B 20] .

4.14. Use the method of Sec.4.5 to construct a multiplier function for $G(s) = s^2/(s^4+3s^3+4s^2+2s+1)$, and hence state the conditions for the exponential stability of (2.6) with $\varphi(\cdot)\in C$, C_m, C_{mo}.

4.15. Repeat Exercise 4.14 for $G(s) = ((2s+1)/s(s+1)) + 0.75 + 0.5e^{-s}$. Use the transformation of Exercise 2.1, if necessary.

4.16. Under what conditions on $G(j\omega)$ and $Z_g(j\omega)$, the latter being defined as in Lemma 4.36, is the functional

$$I = \int_Q^T (g(t) * k(t)\varphi_T(\sigma(t)))(Z_g\sigma_T)(t)dt$$

for $\varphi(\cdot)\in C_m, C_{mo}$ nonnegative. Hence, in the place of the real part condition (hypothesis (i)) of Theorem 4.41, can one write an equivalent time domain condition ? Derive, if possible, on the same lines, stability conditions corresponding to Theorem 4.49 for a periodic $k(t)$.

4.17. Establish conditions under which functional I_{no} of Lemma 4.32 is nonnegative where α and β are now treated as functions of time. What class of feedback systems can be analysed for stability using this extended lemma ?

Derive similar extensions for lemmas 4.33–4.39 and for the corresponding theorems of Table 4.2.

4.18. Consider a nonlinear functional differential equation of the form

$$(du/dt) = -\int_0^t g(t-\tau)\varphi(u(\tau))d\tau + f(t,u(t)) \qquad (4.107)$$

where $u(t)$, $\varphi(\cdot)$, and $f(t,\cdot)$ are continuous scalar functions.

We are interested in the boundedness and asymptotic stability of solutions of this equation. The results of MacCamy and Wong [M 1] apply to the situation when $g(t)$ has a Laplace transform which is positive real. Suppose this condition is not satisfied, extend the multiplier idea to the equation (4.107) to derive (exponential) stability conditions more general than those of [M 1].

4.19. On the basis of a particular nonlinear function $\varphi(\cdot)$, what sort of a physical interpretation can one adduce for the term $(\delta_s-\delta_i)$ which appears in Lemma 4.36 and in other lemmas on the nonnegativeness of a nonlinear time varying gain in cascade with a linear time invariant system ?

CHAPTER 5

L_2-STABILITY OF FEEDBACK SYSTEMS IN GENERAL INTEGRAL FORM

5.1. Introduction. The generalization, achieved in Chapter 4 vis-a-vis the results of Chapter 3, is of two kinds :

(i) The stability conditions of Chapter 4 include those of Chapter 3 as a special case.

(ii) The stability constraint on $k(t)$ in Chapter 4, for the case of $\varphi(\cdot) \in C_m$, takes the form of simultaneous lower and upper global bounds on $\theta(t)$, the normalized rate of variation of $k(t)$. The lower global bound on $\theta(t)$ can be traded for the upper global bound . In contrast with this, the stability constraint on $k(t)$ in Chapter 3, for the case of $\varphi(\cdot) \in C_m$, is either an upper or a lower global bound on $\theta(t)$, a combination of the two bounds being as yet not derivable in the Lyapunov framework.

The generalization of the present chapter is of a different kind. We consider the feedback system

$$\left.\begin{aligned} v(t) &= f(t)-k(t)\varphi(\sigma(t)) \\ \sigma(t) &= (\mathcal{G}_{1} v)(t) \triangleq \sum_{i=1}^{\infty} g_i v(t-\tau_i) + \int_0^{\infty} g(\tau)\, v(t-\tau)d\tau \end{aligned}\right\} t \in [0,\infty) \quad (2.7)$$

where $k(t)$, $\varphi(\cdot)$, g_i's for all i, and $g(\cdot)$ are as defined in Sec.2.2(a) and Sec.2.2(c). We propose to derive L_2-stability conditions (Definition 2.38, page 29) under Assumption S2 of Chapter 2 (page 31) . That is, we deal with Problem 2.41 (page 32). The Laplace transform of the forward time invariant block is defined by

$$G_1(s) = \sum_{i=1}^{\infty} g_i \exp(-s\tau_i) + \int_0^{\infty} g(t)\exp(-st)dt \quad (2.18)$$

The main difference between the stability definitions for (2.7) and those for

$$p(D)y + k(t)\varphi(q(D)y) = 0, \quad t \in [0,\infty) \quad (2.1)$$

lies in the fact that L_2-stability for (2.7) concerns input-output behaviour and hence details of internal structure of the system are omitted, whereas exponential stability for (2.1) refers to the evolution of its state, and consequently the internal structure of the system plays a decisive role. Given an arbitrary system, there is a possibility of having states that are not accessible from the input and not obsevable from the output -- a possibility that is excluded in the L_2-stability theory. However, under suitable conditions, which are satisfied (Sec.2.2,Assumption 2.21, page 23) for the linear system

$$p(D)y + k(t)q(D)y = f(t), \quad t \in [0,\infty) \tag{2.8}$$

with $f(\cdot)$ as the input and $y(\cdot)$ as the output, the two stability conditions are equivalent [A 4] . See Sec.2.41.

Note that the system (2.7) is a slight generalization of the system

$$\sigma(t) = f_o(t) - \int_{t_o}^{t} g(t-\tau)k(\tau)\varphi(\sigma(\tau))d\tau \tag{2.6}$$

considered in Chapter 4. See Sec.2.2(b) and Sec.2.2(c)(Page 25).

The L_2-stability analysis of (2.7) is carried out in the framework of the passive (Hilbert space) operator theory of Zames for which the best reference is the original set of papers [Z 3(a)(b)] . However, in view of the special convolution form of the system (2.7), the simpler 'energy balance' argument is employed to derive the major result (Theorem 5.31) of the chapter. The proof of this result is based on two lemmas (Sec.5.3) : the first lemma (Lemma 5.31) concerns the passivity of two operators, one of which is linear time invariant and the other nonlinear time varying ; the second lemma (Lemma 5.32) deals with the time-varying-gain factorization. These lemmas, which are believed to be of interest in their own right, along with Schwarz's inequality (Lemma 5.33) and the Parseval theorem (Lemma 4.41) are used to complete the proof of Theorem 5.31.

Other results (of the type found in Chapter 4) can be established along similar lines, and hence are omitted. Table 5.1 traces briefly the generalization we have scored from the results of Chapter 3 to those of the present one. It is but natural to attempt a unification of these results, and this has already been achieved by Zames [Z 3(b)] in the well known passivity L_2-stability theorem (Sec.5.4, Theorem 5.41) applicable to operator feedback systems, Fig.5.1, wherein $\mathcal{G}_1$ and $\mathcal{G}_2$ are two arbitrary (causal) operators. On the one hand, it is remarkable that the most general result (Theorem 5.31) known for the L_2-stability of a specific system can be imbedded in Theorem 5.41. On the other hand, it is a little disappointing that no results better than Theorem 5.31 have been derived from Theorem 5.41.

Table 5.1

No.	Feedback system	Type of stability	Stability problem is reduced to establishment of or is based on	Nature of the stability conditions
1	(2.21) Differential form. Linear.	Exponential	first order differential inequality	Nyquist stability requirement on $G(j\omega-\beta)$ for some constant $\beta>0$; $\theta(t)$ to obey an upper or lower global bound involving β.
2	(2.1) Differential form Nonlinear.	Exponential	first order differential inequality	(i) $\varphi(\cdot)\in C_m, C_{mo}$. Multiplier function form. Positive real multiplier function with a time domain constraint ; $\theta(t)$ to obey an upper or lower global bound dependent on the multiplier function and δ_s. (ii) $\varphi(\cdot)\in C$. Popov's criterion to be satisfied by $G(j\omega-\beta)$ for some constant $\beta>0$; $\theta(t)$ to obey an upper or lower global bound dependent on δ_s. (iii) $\varphi(\cdot)\in C$. Re $G(j\omega)\geqslant 0$ for all real ω (special version of the circle criterion).
3	(2.22) Integral form. Linear.	Exponential	first order integral inequality	Multiplier function form. Causal + anticausal multiplier function. Simultaneous upper and lower global bounds on $\theta(t)$ dependent on the multiplier function. Nyquist type geometric condition unknown.
4	(2.6) Integral form. Nonlinear.	Exponential	first order integral inequality	(i) $\varphi(\cdot)\in C_m$, C_{mo}. Multiplier function form. Causal + anticausal multiplier function satisfying a time domain constraint; simultaneous upper and lower global bounds on $\theta(t)$ dependent on the multiplier function (and the nonlinearity $\varphi(\cdot)$). (ii) $\varphi(\cdot)\in C$. See item No.2 (ii) and (iii) above. (iii) $\varphi(\cdot)\in C$. Exercise 5.4.

Fig 5.1. An operator feedback system with single input.

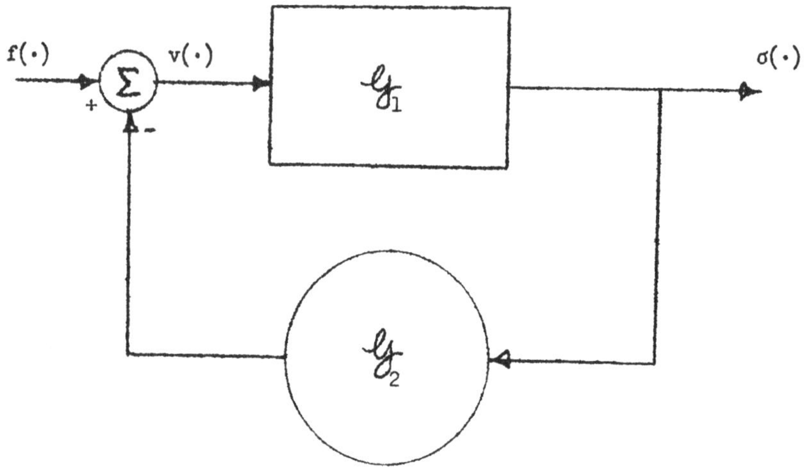

Fig.5.2. An operator feedback system with two inputs.
(Exercise 5.1)

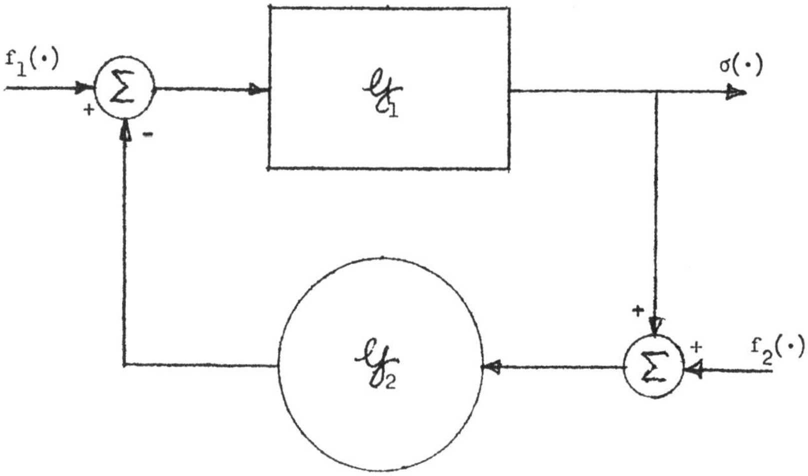

No.	Feedback system	Type of stability	Stability problem is reduced to establishment of or is based on	Nature of the stability conditions
5	(2.7) General integral form : (i) Linear (ii) non-linear	L_2-stability	Schwarz's inequality	(i) See item No.3 above. (ii) See items No.4 (i) and No.2(iii) above.
6	Fig.5.1	L_2-stability	Schwarz's inequality	Passivity requirements on $\mathcal{G}_1$ and $\mathcal{G}_2$; strict passivity and finite gain of one of them. Result due to Zames. [Z 3(b)]
7	(2.7) General integral form with k(t) unbounded as in Acker [A1].	L_2-stability	-----	Exercise 5.5.

The 'transition' from Theorem 5.31 to Theorem 5.41 (or vice versa) requires a lemma on the factorization of operators (Lemma 5.42) on a Hilbert space due to Gohberg and Krein [G 4] . The background needed for a rigorous treatment of the latter is considerable and is hence omitted. However, an intuitive explanation (or statement) of the relevant ideas is given. A major portion of the exercises at the end is primarily meant for the mathematically minded reader.

5.2. Preliminaries. We summarize here only the essential ideas needed in the proof of the main result. For comments on existence and uniqueness of solutions and their role in L_2-stability analysis, see Sec.2.53 and also Haddad [H 2(b) p 1081] .

$L_2(0,\infty)$ is the linear space of real valued functions $x(\cdot)$ on $[0,\infty)$ with the property that

$$\int_0^{\infty} |x(t)|^2 \, dt < \infty .$$

Let $L_2(0,\infty)$ be normed with the norm

$$\|x(\cdot)\| = \left(\int_0^\infty |x(t)|^2 \; dt \right)^{1/2} .$$

Popov introduced the truncated function in order to ensure Fourier transformability of the system solution. Here the idea of extended spaces is used not only to give a suitable definition of stability but also to ensure Fourier transformability of the system solution.

For any real valued function $x(\cdot)$ on $[0,\infty)$ and any $T > 0$, let $x_T(\cdot)$ denote the truncated function defined by

$$x_T(t) = \begin{cases} x(t) & \text{for } t \leqslant T \\ 0 & \text{for } t > T. \end{cases}$$

Definition 5.21. Let L_{2e} be the space of those real valued functions $x(\cdot)$ on $[0,\infty)$ whose truncations $x_T(\cdot)$ belong to $L_2(0,\infty)$ for all $T > 0$.

If $x(\cdot)$ is an element of L_{2e}, then the extended norm of $x(\cdot)$, in symbols $\|x(\cdot)\|_e$, is given by

$$\|x(\cdot)\|_e = \begin{cases} \|x(\cdot)\| & \text{if } x(\cdot) \in L_2 \\ \infty & \text{if } x(\cdot) \notin L_2. \end{cases}$$

Assumption 5.21. The response σ and the error signal v satisfy the conditions $\sigma(\cdot) \in L_{2e}$ and $v(\cdot) \in L_{2e}$ for each input $f(\cdot) \in L_2(0,\infty)$.

Definition 2.38. The system (2.7) is L_2-stable if, for all $f(\cdot) \in L_2(0,\infty)$, $\sigma(\cdot)$ (or $v(\cdot)$) is also in $L_2(0,\infty)$ and

$$\|\sigma\| \leqslant \text{const.} \|f\| \quad \text{or} \quad \|v\| \leqslant \text{const.} \|f\|.$$

Assumption 5.22. Further assumptions made on the constituents of (2.7) are as follows (See Sec.2.32 and Sec.2.2(c)) : $g(\cdot)$ is an element of $L_1(0,\infty)$, i.e.,

$$\int_0^\infty |g(t)| \, dt < \infty ,$$

and there is a constant $\eta_0 > 0$ such that $g(t)\exp(\eta_0 t)$ is also in $L_1(0,\infty)$; $\{g_i\}$ is a sequence in ℓ_1, i.e., $\sum_i |g_i| < \infty$, and $\{\tau_i\}$ is a sequence in $[0,\infty)$. For the assumptions on $k(\cdot)$ and $\varphi(\cdot)$, see Sec.2.2(a).

Let $G_1(j\omega)$ denote the frequency function of the linear time invariant part of the feedback system, i.e.,

$$G_1(j\omega) = \sum_{i=1} g_i \exp(-j\,\omega\tau_i) + \int_0^{\infty} g(t)\exp(-j\omega\, t)dt. \tag{5.1}$$

The L_2-stability problem (PROBLEM 2.41) can be stated as follows: Find conditions on $k(t)$ and $G_1(j\omega)$ which ensure that the feedback system (2.7) with $\varphi(\cdot) \in C$, C_m or C_{mo} is L_2-stable.

A solution to the stability problem involves an additional definition.

Definition 5.22. Let $\mathcal{O}$ denote the class of operators $\mathcal{Z} : L_{2e} \to L_{2e}$ satisfying an equation of the type

$$(\mathcal{Z}x)(t) = x(t) + \sum_{i=1}^{\infty} (z_i x(t-\mu_i) + z_i' \, x(t+\mu_i')) + \int_{-\infty}^{\infty} z(\tau)x(t-\tau)d\tau \tag{5.2}$$

where the sequences $\{z_i\}$ and $\{z_i'\}$ are in ℓ_1, i.e., $\sum_{i=1}^{\infty}(|z_i| + |z_i'|)<\infty$; sequences $\{\mu_i\}$, $\{\mu_i'\}$ are in $[0,\infty)$; $z(\cdot)$ is a real valued function on $(-\infty,\infty)$, and is in $L_1(-\infty,\infty)$, i.e., $\int_{-\infty}^{\infty}|z(t)|\,dt<\infty$. The frequency function $Z(j\omega)$ is given by

$$Z(j\omega) = 1 + \sum_{i=1}^{\infty} (z_i\exp(-j\,\omega\,\mu_i) + z_i'\exp(j\omega\,\mu_i)) + \int_{-\infty}^{\infty} z(\tau)\exp(-j\,\omega\tau)d\tau$$

We recall the following notations used in Chapters 3 and 4.

$$\text{With } \theta(t) = (\tfrac{dk}{dt}/k),$$

$$\begin{aligned}\theta^+(t) &= \theta(t) \text{ for all } \theta(t) > 0\\ &= 0 \quad \text{for all } \theta(t) \leq 0 \text{ ; and}\\ \theta^-(t) &= \theta(t) \text{ for all } \theta(t) < 0\\ &= 0 \quad \text{for all } \theta(t) \geq 0.\end{aligned}$$

Evidently, $\theta(t) = \theta^+(t) + \theta^-(t)$.

Further,

$$\begin{aligned}\delta_s &= \sup_{\substack{\sigma\\ \sigma\neq 0}} \left(\int_0^{\sigma} \varphi(w)dw/\varphi(\sigma)\sigma\right)\\ \delta_i &= \inf_{\substack{\sigma\\ \sigma\neq 0}} \left(\int_0^{\sigma} \varphi(w)dw/\varphi(\sigma)\sigma\right)\end{aligned} \tag{5.3}$$

Note that for $\varphi(\cdot)\in C_m$, $0 < \delta_s \leq 1$, and for $\varphi(\cdot)\in C$, $0 < \delta_s \leq \infty$.

5.3. Main Result and its Proof.

The following theorem, Theorem 5.31, resembling Theorem 4.41 in combination with Lemma 4.43, is the main result. Other results similar to those listed in Table 4.2 can be established on the lines of the proof of Theorem 5.31.

Theorem 5.31. The feedback system (2.7) with $\varphi(\cdot)\in C_m$ is L_2-stable if there exists an operator Z in $\mathcal{O}$ with $z(\cdot) \leq 0$, $z_i \leq 0$ and $z_i' \leq 0$ for all i =1,2, 3,$\cdots$, such that

(a) Re $Z(j\omega - \varepsilon_o)G_1(j\omega - \varepsilon_o) \geq \delta > 0$ for some positive constants ε_o and δ (with ε_o, by virtue of Assumption 5.22, less than η_o), and for all real ω;

(b) for some positive constants γ_c and γ_{ac}

$$\sum_{i=1}^{\infty}|z_i| \exp(\gamma_c\mu_i) + \sum_{i=1}^{\infty}|z_i'| \exp(\gamma_{ac}\mu_i') + \int_0^{\infty}|z(\tau)| \exp(\gamma_c\tau)d\tau$$

$$+\int_{-\infty}^{0}|z(\tau)| \exp(-\gamma_{ac}\tau)d\tau \leq 1/(1+\delta_s-\delta_i) \; ; \tag{5.4}$$

and (c) for some positive constants N_1 and N_2, and for all finite $T > 0$ and all $t_o \geq 0$,

$$(1/T)\int_{t_o}^{t_o+T}\theta^+(t)dt \leq N_1 \; ; \quad -N_2 \leq (1/T)\int_{t_o}^{t_o+T}\theta^-(t)dt$$

and in addition one of the following sets of inequalities is satisfied :

Set 1. (i) $\lim_{T\to\infty} (1/T)\int_{t_o}^{t_o+T}\theta^+(t)dt \leq \gamma_c$

and (ii) $\lim_{T\to\infty} (1/T)\int_{t_o}^{t_o+T}\theta^-(t)dt \geq -\gamma_{ac}$.

Set 2. (i) For $\gamma_c < \gamma_{ac}$,

$$\lim_{T\to\infty} (1/T)\int_{t_o}^{t_o+T}\theta^-(t)dt \geq \gamma_c - \gamma_{ac}, \text{ and}$$

$$\lim_{T\to\infty} (1/T)\int_{t_o}^{t_o+T}\theta^+(t)dt = 0.$$

(ii) For $\gamma_c > \gamma_{ac}$,

$$\lim_{T\to\infty} (1/T) \int_{t_o}^{t_o+T} \theta^+(t)dt \leq \gamma_c - \gamma_{ac}, \text{ and}$$

$$\lim_{T\to\infty} (1/T) \int_{t_o}^{t_o+T} \theta^-(t)dt = 0.$$

(iii) For $\gamma_c = \gamma_{ac}$, $\theta(t)$ is unrestricted.

The proof of Theorem 5.31 is based on two lemmas : the first lemma (Lemma 5.31) concerns the positivity of two operators in cascade, one of them time invariant and the other nonlinear time varying ; the second lemma (Lemma 5.32) deals with the factorization of time varying gains.

Let $\mathcal{K}$ be the class of absolutely continuous real valued functions $k(\cdot)$ on $[0,\infty)$ with each $k(\cdot)$ having constants $\underline{k} > 0$ and $\bar{k} \geq \underline{k}$ for which $\underline{k} \leq k(t) \leq \bar{k}$ for all $t \geq 0$.

Lemma 5.31. If (a) the operator $\mathcal{Z}$ belongs to $\mathcal{O}$ with $z(\cdot) \leq 0$, $z_i \leq 0$ and $z_i' \leq 0$ for all $i = 1,2,3,\cdots$, (b) for some nonnegative constants γ_c and γ_{ac}, inequality (5.4) is satisfied; and (c) for these values of γ_c and γ_{ac}, with $h(\cdot) \in \mathcal{K}$, $h(t)k(t)\exp(-\gamma_c t)$ is nonincreasing and $h(t)k(t)\exp(\gamma_{ac} t)$ nondecreasing for all $t \geq 0$, then the following inequality holds :

$$\int_0^T h(t)k(t)(\mathcal{Z}x)(t)\varphi(x(t))dt \geq 0 \tag{5.5}$$

for $\varphi(\cdot) \in C_m$, for all x in the domain of $\mathcal{Z}$ and for all $T \geq 0$.

Proof of Lemma 5.31. See the proof of Lemma 4.36 in Appendix 4.2.

Remark 5.31. For $\varphi(\cdot) \in C_{mo}$, Lemma 5.31 holds without, in hypothesis (a) of Lemma 5.31, the nonpositivity constraint on $z(\cdot)$, z_i and z_i' for all $i = 1,2,3,\cdots$.

Lemma 5.32. If there exists a time-multiplier function $h(\cdot)$ in $\mathcal{K}$ satisfying hypothesis (c) of Lemma 5.31 for some positive constants γ_c and γ_{ac}, then hypothesis (c) of Theorem 5.31 holds.

Proof of Lemma 5.32. See the proof of Lemma 4.43 in Appendix 4.5.

Lemma 5.33 (Schwarz). If $f_1(\cdot)$ and $f_2(\cdot)$ belong to $L_2(0,\infty)$, then $f_1(\cdot)f_2(\cdot)$

belongs to $L_1(0,\infty)$ and

$$\left| \int_0^\infty f_1(t)f_2(t)dt \right| \leq \left(\int_0^\infty f_1^2(t)dt\right)^{1/2} \left(\int_0^\infty f_2^2(t)dt\right)^{1/2}.$$

The above lemma is well known (but its converse, used in Chapter 7, is not).

Based on Lemmas 5.31 and 5.32 (in combination with Lemmas 5.33 and 4.41), Theorem 5.31 may now be proved.

Proof of Theorem 5.31. Consider the integral, for any $T > 0$,

$$\rho(T) = \int_0^T h(t)f(t)(z\,\mathcal{G}_1 v)(t)dt \tag{5.6}$$

which on employing (2.7) becomes

$$\rho(T) = \int_0^T h(t)\,v(t)(z\,\mathcal{G}_1 v)(t)dt + \int_0^T h(t)k(t)\varphi(\sigma(t))(z\sigma)(t)dt \tag{5.7}$$

The first integral on the right hand side of (5.7) is, for some $\varepsilon > 0$,

$$\int_0^T h(t)\,v(t)(z\,\mathcal{G}_1 v)(t)dt = \int_0^T h(t)\exp(-2\varepsilon t)\,v(t)(z\,\mathcal{G}_1 v)(t)\exp(2\varepsilon t)dt.$$

Suppose we choose $h \in \mathcal{K}$ so that $h(t)\exp(-2\varepsilon t)$ is nonincreasing and $\varepsilon < \varepsilon_0$ (see hypothesis (a) of Theorem 5.31). Then we can invoke the second mean value theorem (Lemma 4.31), according to which there is a point T' in $[0,T]$ such that

$$\int_0^T h(t)v(t)(z\,\mathcal{G}_1 v)(t)dt = h(0)\int_0^{T'} v(t)(z\,\mathcal{G}_1 v)(t)\exp(2\varepsilon t)dt$$

where $h(0) > 0$ in view of our choice of $h \in \mathcal{K}$. By virtue of hypothesis (a) of the theorem statement, $\mathrm{Re}\, Z(j\omega-\varepsilon_0)G_1(j\omega-\varepsilon_0) \geq \delta > 0$ for all ω in $(-\infty,\infty)$. Hence by Parseval's theorem (Lemma 4.41),

$$\int_0^T h(t)v(t)(z\,\mathcal{G}_1 v)(t)dt \geq \delta_1 \int_0^T v^2(t)\exp(2\varepsilon t)dt \tag{5.8}$$

for some constant $\delta_1 > 0$.

Now consider the second integral on the right hand side of (5.7). By virtue of Lemmas 5.31 and 5.32, this integral is nonnegative if hypotheses (b) and (c) of the theorem **statement** are satisfied. Hence $\rho(T)$ as defined by

(5.6) satisfies the inequality

$$\delta_1 \int_0^T v^2(t)\exp(2\varepsilon t)dt \leq \rho(T) = \int_0^T h(t)f(t)(z\,\mathcal{G}_1 v)(t)dt \tag{5.9}$$

for some constant $\delta_1 > 0$.

As regards the right hand side of (5.9), we have, in view of the assumption that $h \in \mathcal{K}$,

$$\int_0^T h(t)f(t)(z\,\mathcal{G}_1 v)(t)dt = h(0)\int_0^{T'} f(t)(z\,\mathcal{G}_1 v)(t)\exp(2\varepsilon_1 t)dt \tag{5.10}$$

for some $\varepsilon_1 > 0$ and T' in $[0,T]$.

Invoking Parseval's theorem, (5.10) can be rewritten as

$$\int_0^T h(t)f(t)(z\,\mathcal{G}_1 v)(t)dt = (h(0)/2\pi)\int_{-\infty}^{\infty} F_{T'}(-j\omega-\varepsilon_1)Z(j\omega-\varepsilon_1)G_1(j\omega-\varepsilon_1) V_{T'}(j\omega-\varepsilon_1)d\omega \tag{5.11}$$

where $F_{T'}(j\omega)$ and $V_{T'}(j\omega)$ are respectively the Fourier transforms of $f_{T'}(t)$ and $v_{T'}(t)$.

From (5.11), we conclude that

$$\int_0^T h(t)f(t)(z\,\mathcal{G}_1 v)(t)dt \leq (h(0)/2\pi) \sup_{-\infty<\omega<\infty} |Z(j\omega-\varepsilon_1)G_1(j\omega-\varepsilon_1)| \left|\int_{-\infty}^{\infty} F_{T'}(-j\omega-\varepsilon_1) V_{T'}(j\omega-\varepsilon_1)d\omega\right|$$

from which we have

$$\int_0^T h(t)f(t)(z\,\mathcal{G}_1 v)(t)dt \leq h(0)\sup_{-\infty<\omega<\infty} |Z(j\omega-\varepsilon_1)G_1(j\omega-\varepsilon_1)| \left|\int_0^{T'} \exp(2\varepsilon_1 t) v(t)f(t)dt\right| \tag{5.12}$$

Noting that $0 \leq T' \leq T$, apply Schwarz's inequality to the right hand integral of (5.12) to get

$$\int_0^T h(t)f(t)(z\,\mathcal{G}_1 v)(t)dt \leq h(0)\sup_{-\infty<\omega<\infty} |Z(j\omega-\varepsilon_1)G_1(j\omega-\varepsilon_1)| \left(\int_0^T v^2(t)\exp(4\varepsilon_1 t)dt\right)^{1/2}\left(\int_0^T f^2(t)dt\right)^{1/2} \tag{5.13}$$

By choosing $\varepsilon_1 = \varepsilon/2$ where $\varepsilon < \varepsilon_0$ of hypothesis (a) of the theorem statement, and noting that $\sup_{-\infty<\omega<\infty} \left| Z(j\omega - \varepsilon_1)G_1(j\omega - \varepsilon_1)\right|$ is finite by virtue of the assumptions on $\mathcal{Z}$ and $\mathcal{G}_1$, inequality (5.13) reduces to

$$\int_0^T h(t)f(t)(\mathcal{Z}\,\mathcal{G}_1 v)(t)dt \leq A\left(\int_0^T v^2(t)\exp(2\varepsilon t)\,dt\right)^{1/2}\left(\int_0^T f^2(t)dt\right)^{1/2} \qquad (5.14)$$

where $A = h(0) \sup_{-\infty<\omega<\infty} \left| Z(j\omega - (\varepsilon/2))G_1(j\omega - (\varepsilon/2))\right|$.

Inequality (5.14) in combination with the inequality (5.9) yields

$$\delta_1\left(\int_0^T v^2(t)\exp(2\varepsilon t)dt \leq A\left(\int_0^T v^2(t)\exp(2\varepsilon t)dt\right)^{1/2}\left(\int_0^T f^2(t)dt\right)^{1/2}$$

from which we infer that

$$\delta_1\left(\int_0^T v^2(t)\exp(2\varepsilon t)dt\right)^{1/2} \leq A\left(\int_0^T f^2(t)dt\right)^{1/2} \qquad (5.15)$$

which is valid for all $T > 0$. Recalling that $0 < \varepsilon < \varepsilon_0$ of hypothesis (a) of the theorem statement, we conclude that the feedback system (2.7) with $\varphi(\cdot) \in C_m$ is L_2-stable. The theorem is proved.

Corollary 5.311. Theorem 5.31 applies to the feedback system (2.7) with $\varphi(\cdot) \in C_{mo}$ under less restrictive assumptions on the operator $\mathcal{Z}$ in $\mathcal{O}$: the non-positivity constraint on $z(\cdot), z_i$ and z'_i is no longer necessary. (See Remark 5.31).

We now merely state the theorem for the linear feedback system (2.23) which is obtained from (2.7) by substituting $\sigma(t)$ for $\varphi(\sigma(t))$. The proof of the theorem is left as an exercise. (See, however, [S 14(c)]).

Let $z_c(t) = z(t)$ for $t \geq 0$; $z_{ac}(t) = z(t)$ for $t \leq 0$
$\quad\quad = 0$ for $t < 0$; $\quad 0$ for $t > 0$,

where $z(\cdot)$ refers to the function in (5.2) defining the operator $\mathcal{Z}$ in $\mathcal{O}$. Further, let $Z_c(j\omega)$ and $Z_{ac}(j\omega)$ be the Fourier transforms of $z_c(t)$ and $z_{ac}(t)$ respectively.

Theorem 5.32. The linear feedback system (2.23) is L_2-stable if there exists an operator $\mathcal{Z}$ in $\mathcal{O}$ such that

(a) Re $Z(j\omega-\varepsilon_o)G_1(j\omega-\varepsilon_o) \geq \delta > 0$ for some positive constants ε_o and δ (where ε_o, by virtue of Assumption 5.22, is less than η_o), and for all real ω;

(b) for some positive constants $\gamma_{c\ell}$ and $\gamma_{ac\ell}$,

$$\begin{aligned} &\mathrm{Re}\left\{\sum_i z_i \exp((-j\omega-\gamma_{c\ell})\mu_i) + Z_c(j\omega-\gamma_{c\ell})\right\} \geq 0, \\ &\mathrm{Re}\left\{\sum_i z_i' \exp((j\omega+\gamma_{ac\ell})\mu_i') + Z_{ac}(j\omega+\gamma_{ac\ell})\right\} \geq 0 \end{aligned} \tag{5.16}$$

for all real ω;

and (c) hypothesis (c) of Theorem 5.31 is satisfied with γ_c and γ_{ac} replaced by $\gamma_{a\ell}$ and $\gamma_{ac\ell}$ respectively.

Remark 5.32. Suppose Theorem 5.31 were applied to the linear feedback system (2.23). Then $\delta_s = \delta_i = (1/2)$ from the defining relations (5.3) and hence, in hypothesis (b) of Theorem 5.31, the right hand side of the time-domain inequality (5.4) concerning the operator $\mathcal{Z}$ would be merely 1. This inequality takes the place of hypothesis (b), a frequency domain inequality, of Theorem 5.32 where there is no time-domain bound on the operator $\mathcal{Z}$.

Remark 5.33. For a geometric interpretation of Theorem 5.31, see Sec.4.5 which needs little modification. In view of Remark 5.32, Theorem 5.31 needs improvement.

5.4. Generalization to Feedback Systems in Operator Form.

The wording of lemmas and theorems in Sec.5.3 is such as to pave the way for an operator-theoretic statement of L_2-stability results concerning the feedback system of Fig.5.1 where $\mathcal{G}_1$ and $\mathcal{G}_2$ are general (causal)* operators. Before stating the main results of this section, we need a few more definitions [Z 4].

Let the operator $\mathcal{H}$ be a mapping of L_2 into L_2 (or of L_{2e} into L_{2e}).

Definition 5.41. $\mathcal{H}$ is nonanticipative or causal if

$$(\mathcal{H}x)_T(t) = (\mathcal{H}x_T)_T(t)$$

for all x in L_2 (or L_{2e}) and all t in $[0,\infty)$.

Definition 5.42. Let $\mathcal{H}$ be a nonanticipative mapping of L_{2e} into L_{2e}. Then the gain of $\mathcal{H}$, $\gamma_e(\mathcal{H})$, is given by

---

* See Definition 5.41 below.

$$\gamma_e(\mathcal{H}) = \sup_{\substack{x \in L_{2e} \\ T \in [0,\infty) \\ x_T \neq 0}} \left\{ \|(\mathcal{H}x(\cdot))_T\|_e \,/\, \|x_T(\cdot)\|_e \right\}.$$

$\mathcal{H}$ is said to be passive if

$$\int_0^\infty x_T(t)(\mathcal{H}x)_T(t)dt \geq 0$$

for all $x(\cdot)$ in L_{2e} and all T in $[0,\infty)$.

$\mathcal{H}$ is said to be 'strongly passive' if

$$\int_0^\infty x_T(t)(\mathcal{H}x)_T(t)dt \geq \delta \|x_T(\cdot)\|_e^2$$

for all $x(\cdot)$ in L_{2e}, all T in $[0,\infty)$ and some $\delta > 0$.

A mapping $\mathcal{H}^*$ of L_{2e} into L_{2e} is said to be an adjoint of $\mathcal{H}$ if

$$\int_0^\infty x_T(t)(\mathcal{H}y_T)(t)dt = \int_0^\infty (\mathcal{H}^* x_T)(t)y_T(t)dt$$

for all $x(\cdot)$, $y(\cdot)$ in L_{2e} and all T in $[0,\infty)$.

See Exercise 5.6 concerning the nonanticipative and anticipative properties of $\mathcal{H}$ and $\mathcal{H}^*$.

The operator $\mathcal{H}$ is said to be invertible if $\mathcal{H}$ has an inverse, i.e., there exists an element $\mathcal{H}^{-1}$ such that $\mathcal{H}^{-1}\mathcal{H} = \mathcal{H}\mathcal{H}^{-1} = E$, the unit operator.

5.41. We now consider the feedback system (Fig.5.1) described by the set of equations

$$\begin{aligned} v(\cdot) &= f(\cdot) - \mathcal{G}_2\sigma(\cdot) \\ \sigma(\cdot) &= \mathcal{G}_1 v(\cdot). \end{aligned} \tag{5.17}$$

The assumptions on $v(\cdot)$, $f(\cdot)$ and $\sigma(\cdot)$ are the same as before (Sec.5.2, Assumption 5.21). We wish to determine conditions which ensure that $\sigma(\cdot)$ and $v(\cdot)$ are in L_2.

Theorem 5.41 (Zames [Z 3(b)]). If $\mathcal{G}_1$ and $\mathcal{G}_2$ are passive and if either $\mathcal{G}_1$ is strongly passive and $\gamma_e(\mathcal{G}_1)$ is finite or $\mathcal{G}_2$ is strongly passive and $\gamma_e(\mathcal{G}_2)$ is finite, then $\sigma(\cdot)$ and $v(\cdot)$ are in L_2.

Proof. Consider the integral, for any $T > 0$,

$$\rho_1(T) = \int_0^T f(t)(\mathcal{G}_1 v)(t)dt \tag{5.18}$$

which on employing (5.17) becomes

$$\rho_1(T) = \int_0^T v(t)(\mathcal{G}_1 v(t)dt + \int_0^T \sigma(t)(\mathcal{G}_2\sigma)(t)dt. \tag{5.19}$$

Suppose $\mathcal{G}_2$ is passive, $\mathcal{G}_1$ is strictly passive and has a finite gain. Then from (5.19) and (5.18), invoking Schwarz's inequality,

$$\delta_1 \| v_T \|^2 \leq \rho_1(T) \leq \gamma_e(\mathcal{G}_1) \| f_T \| \cdot \| v_T \| \tag{5.20}$$

for some positive constant δ_1 and for all $T > 0$. From inequality (5.20), there follows

$$\| v_T \| \leq (\gamma_e(\mathcal{G}_1)/ \delta_1) \| f_T \|$$

for all $T > 0$. Thus $v(\cdot)$ is in L_2. Furthermore from the second equation of (5.17), $\sigma(\cdot) = (\mathcal{G}_1 v)(\cdot)$ from which $\|\sigma(\cdot)\|_e \leq \gamma_e(\mathcal{G}_1) \|v(\cdot)\|_e$ so that $\sigma(\cdot)$ is also in L_2. The case where $\mathcal{G}_2$ is strongly passive is treated in a similar way by choosing the integral, for any $T > 0$,

$$\rho_2(T) = \int_0^T f(t)(\mathcal{G}_2\sigma)(t)dt.$$

The theorem is proved.

Remark 5.41. Suppose $\mathcal{G}_1$ and $\mathcal{G}_2$ are not passive, then motivated by the results of Chapters 3 and 4, it is natural to construct a noncausal 'multiplier operator' $\mathcal{M}$, which when suitably combined (or 'composed') with $\mathcal{G}_1$ and $\mathcal{G}_2$ gives rise to passive operators. The method of dealing with the noncausal multiplier function as employed in Sec.5.3 (and in Chapter 4) is applicable to time invariant convolution operators in feedback with a time varying nonlinear gain (i.e., referring to Fig.5.1 with $\mathcal{G}_1$ = time invariant convolution operator, $\mathcal{G}_2$ = a time varying nonlinear gain). For the general case of Fig.5.1, we have to resort to different method which is the subject of the following lemma, due to Zames and Falb [Z 4, Theorem 2]. This lemma extends the applicability of Theorem 5.41 to the case where $\mathcal{G}_1$ and $\mathcal{G}_2$ are not passive.

Lemma 5.41. Suppose that there are operators $\mathcal{N}_-$ and $\mathcal{N}_+$ composing the operator $\mathcal{N}$, i.e., $\mathcal{N} = \mathcal{N}_-\mathcal{N}_+$, having the following properties :

(a) (i) $\mathcal{N}_-$ and $\mathcal{N}_+$ are invertible ;

(ii) $\mathcal{N}_+$, $\mathcal{N}_+^{-1}$, $\mathcal{N}_-^*$ and $\mathcal{N}_-^{*-1}$ are nonanticipative, have finite gains $\gamma(\cdot)$;

(b) (i) $\mathcal{N}\mathcal{G}_1$ and $\mathcal{N}^*\mathcal{G}_2$ are passive;

(ii) either $\mathcal{N}\mathcal{G}_1$ is strongly passive and $\gamma(\mathcal{G}_1)$ is finite or $\mathcal{N}^*\mathcal{G}_2$ is strongly passive and $\gamma(\mathcal{G}_2)$ is finite.

Then $\sigma(\cdot)$ and $v(\cdot)$ are in L_2.

Proof. We transform the set of equations (5.18) so as to be able to apply Theorem 5.41.

Let $f_1 = (\mathcal{N}_-^*)f$, $\sigma_1 = (\mathcal{N}_+)\sigma$,

$v_1 = (\mathcal{N}_-^*)v$,

and $\mathcal{G}_1' = (\mathcal{N}_+)\mathcal{G}_1(\mathcal{N}_-^{*-1})$,

$\mathcal{G}_2' = (\mathcal{N}_-^*)\mathcal{G}_2(\mathcal{N}_+^{-1})$.

Then the set of equations (5.18) becomes

$$v_1 = f_1 - \mathcal{G}_2'\sigma_1$$

$$\sigma_1 = \mathcal{G}_1' v_1.$$

Note that $\mathcal{G}_1'$ and $\mathcal{G}_2'$ are, in view of hypothesis (a), nonanticipative.

Now apply Theorem 5.41 to express L_2-stability conditions in terms of (i) the passivity of $\mathcal{G}_1'$ and $\mathcal{G}_2'$; (ii) strict passivity and finite gain of $\mathcal{G}_1'$ or $\mathcal{G}_2'$. The definitions of $\mathcal{G}_1'$ and $\mathcal{G}_2'$, and of the adjoint operator along with the finite gain properties of the operators as given in hypothesis (a) will then imply that hypothesis (b) guarantees the L_2-stability of the feedback system (5.18).

5.42. Factorization of an Operator.

In view of Lemma 5.41, we need to find conditions under which an operator $\mathcal{N}$ can be written as a composition of two operators, $\mathcal{N}_-$ and $\mathcal{N}_+$, satisfying hypothesis (a) of Lemma 5.41. This factorization is a generalization of the well known result of Wiener for time invariant convolution operators already used in Chapter 3 (Sec.3.3) for generating quadratic forms. According to this result, if $u(s)$ is an even polynomial with real coefficients and $\mathrm{Re}\, u(j\omega) \geq 0$ for all real ω, then there exists a unique polynomial $v(s)$ with real positive coefficients such that $v(s)v(-s) = u(s)$ and $v(s) \triangleq (u(s))^{(+)}$ has zeros only in the closed left half plane $\mathrm{Re}\, s \leq 0$. As an extension, if $u_1(s)$ and $u_2(s)$ are two polynomials with real coefficients and $\mathrm{Re}(u_1(j\omega)/u_2(j\omega)) \geq 0$ for all real ω and 'Ev' denotes the 'even part of', then there is a unique factor $(\mathrm{Ev}\, u_1(s)u_2(-s))^{(+)}$.

Lemma 3.22, concerning the (scalar) multiplier function for the time invariant finite dimensional linear feedback system (2.15), asserts the existence of a rational function $M(s)$ such that $\mathrm{Re}\, M(j\omega) \geq 0$ for all real ω. But the resulting factorization of $M(s)$ has been used for stability analysis in an indirect way (namely, for the generation of quadratic forms) in Chapter 3. However, it is known that factorization is basic to the theory Wiener-Hopf scalar and vector integral equations (see the references in Gohberg and Krein [G 4]).

As regards the factorization of the operator $\mathcal{J}$, Gohberg and Krein [G 4] have derived the corresponding conditions. A complete description of the Gohberg and Krein theorem involves the theory of Banach algebras and is hence beyond the scope of the present monograph. However, we mention below some essential points concerning it. The treatment here is definitely incomplete. For further details, see the references in [G 4] .

Definition 5.43 A Banach space $\mathcal{B}$ is said to be a Banach algebra if there is a multiplication defined in $\mathcal{B}$ which satisfies the inequality

$$\|xy\| \leq \|x\| \cdot \|y\| \quad (x \text{ and } y \in \mathcal{B}),$$

the associative law $x(yz) = (xy)z$, the distributive laws

$x(y+z) = xy + xz$, $(y+z)x = yx + zx$ (x y and $z \in \mathcal{B}$) and the relation

$(\alpha x)y = x(\alpha y) = \alpha(xy)$ where α is any scalar.

The algebra is said (i) to be commutative if $xy = yx$, x and $y \in \mathcal{B}$; (ii) to have a unit if there exists an element $e \in \mathcal{B}$ such that $xe = ex = x$ for all $x \in \mathcal{B}$. An element x of $\mathcal{B}$ with a unit e is said to be invertible if x has an inverse in $\mathcal{B}$, i.e., if there exists an element x^{-1}

such that $x^{-1}x = xx^{-1} = e$. If x and y are invertible in $\mathcal{B}$, so are x^{-1} and xy, since $(xy)^{-1} = y^{-1}x^{-1}$.

Example 5.41. The space L_1 is a Banach algebra if we define multiplication of the functions in L_1 by convolution ; since

$$\| f * g \|_1 \leq \| f \|_1 \cdot \| g \|_1$$

the norm inequality is satisfied. The remaining requirements of Definition 5.43 can be verified to hold in L_1. Thus L_1 is a commutative Banach algebra. It should be noted that L_1 has no unit.

Example 5.42. We are considering convolutions of the form (5.2) where $x(\cdot)$ is an element of $L_2(-\infty,\infty)$. Any convolution of the form (5.2) is a bounded linear transformation of $L_2(-\infty,\infty)$ into itself and the set of all such convolutions can be viewed as a commutative Banach algebra $\mathcal{B}$ with an unit element [Z 4, Sec.5].

The factorization of an operator requires the concept of a projection operator which as applied to our problem is simply the truncation operator.

Let $(P_+x)(t) = x_T(t)$

and $(P_-x)(t) = (E - P_+)x(t)$

where E is the unit operator, i.e., $(Ex)(t) = x(t)$ for all x in $L_2(-\infty,\infty)$. It can be verified that

$$P_+P_+ = P_+ \quad \text{and} \quad P_-P_- = P_-$$

P_+ is said to be a natural projection operator on $\mathcal{B}$. Denote the ranges of P_+ and P_- by $\mathcal{B}_+$ and $\mathcal{B}_-$. Let $\widetilde{\mathcal{B}}_+$ be the subspace spanned by $\mathcal{B}_+$ and E, and let $\widetilde{\mathcal{B}}_-$ be the subspace spanned by $\mathcal{B}_-$ and E. The operators in $\widetilde{\mathcal{B}}_+$ are nonanticipative, and the adjoint of any operator in $\widetilde{\mathcal{B}}_-$ lies in $\widetilde{\mathcal{B}}_+$ and is, therefore, nonanticipative. (See Exercise 5.6).

The following factorization lemma, due to Gohberg and Krein [G 4], is more general than the one given in Zames and Falb [Z 4, Sec.4, Lemma 3].

Lemma 5.42. Let $\mathcal{B}$ be a commutative Banach algebra with an identity E and with norm $\gamma(\cdot)$. If $\mathcal{J}$, any nonzero element of $\mathcal{B}$, is passive, then there are elements $\mu_+ \in \tilde{\mathcal{B}}_+$ and $\mu_- \in \tilde{\mathcal{B}}_-$ such that

(i) $\mu_+\mu_- = \mu$; and

(ii) μ_+ and μ_- are invertible and μ_+^{-1} and μ_-^{-1} are in $\tilde{\mathcal{B}}_+$ and $\tilde{\mathcal{B}}_-$ respectively.

Remark 5.42 For clarification regarding the uniqueness problem of factorization and how it is overcome (by introducing the 'special factorization'), see Gohberg Krein [G 4]. Other authors have extended the lemma as found in Zames and Falb [Z 4]. See Sundaresan and Thathachar [S 14(c)] and Willems [W 3(c), Sec. 3.7, p 72].

Remark 5.43. Suppose the requirement in Lemma 5.41 (for the L_2-stability of the feedback system of Fig.5.1) that $\mu_-^* \mathcal{G}_2 \mu_+^{-1}$ be passive is applied to the special case of the feedback system (2.7). The most general result known concerning the passivity of $\mu_-^* \mathcal{G}_2 \mu_+^{-1}$, with $\mathcal{G}_2$ replaced by the nonlinear time varying gain, is Lemma 5.31. This lemma involves a time invariant convolution operator $\mathcal{Z}$ defined by (5.2). No other result, for instance, using a time varying convolution operator is known. (See Exercise 5.10).

5.5. Conclusions.

The L_2-stability analyses of feedback systems described by the specific integral equation (2.7) and by a more general operator equation (5.18) are unified in the well known passive operator theorem of Zames [Z 3(b)]. As applied to the equation (2.7), the main result (Theorem 5.31) resembles that of Chapter 4 (Theorem 4.41 along with Lemma 4.43), the difference being that input-output stability is used here in the place of exponential stability of Chapter 4.

We have not referred to stability results of other authors, like Willems [W 3(b)-(d)] and Damborg [D 1] [D 2], who express their conditions in terms of inverse operators and causality. On the one hand, the role of causality seems to be controversial in view of the remarks of Haddad [H 2, p 1078] and Skoog [S 5, p 84]. On the other hand, these stability conditions are not verifiable independently (for instance, in the case of time invariant linear operators, without reference to the Nyquist-type criteria) even for the integral equation (2.7), let alone for arbitrary feedback systems. It is quite possible that the results of the present chapter (and those of Chapter 4) could be used to establish the existence of inverse operators and causality properties of the inverse operators, if indeed such a characterization is relevant to stability analysis.

In view of the relative slackening of the pace of publications in the area of time varying system stability, we are tempted to conclude, following John von Neumann, that we have reached the baroque stage. But on closer investigation, we discover that all the results available on stability when applied to Mathieu's or Hill's equation, (1.6) or (1.7), are incapable of reproducing the well known classical stability boundaries (Exercises 5.8 and 5.9). Further, only Problem 2.41 (Chapter 2, p 32) has been attempted using the Lyapunov-Corduneanu (Chapter 3), the Popov (Chapter 4) and the Zames-Sandberg (the present chapter) methods. No results are known for Problem 2.43 (Chapter 2, p 34).

We end the treatment of stability analysis with the following definition of a theorist from Physicists Continue to Laugh, MIR Publishing House, Moscow 1968 : "When a theoretical physicist is asked, let us say, to calculate the stability of an ordinary four-legged table, he rapidly enough arrives at preliminary results which pertain to a one legged table or a table with an infinite number of legs. He will spend the rest of his life unsuccessfully solving the ordinary problem of the table with an arbitrary finite number of legs."

The reader should now be able to recognize that the definition fits a stability theorist to a T with just a minor change of words : system for chair, first order for single leg and second order for four legged.

As regards instability analysis, which is the subject of the next two chapters (in an ascending order of generality adopted for stability analysis), the same observation doubtless holds. In fact, the results available on instability of the feedback system (2.1) or (2.7) have not, in the literature, reached a definitive stage.

EXERCISES

5.1. Suppose the feedback system of Fig 5.1 has an additional input point as shown in Fig.5.2. Are the L_2-stability theorems of the chapter applicable to this system ?

5.2. In engineering applications, it is often more interesting and relevant to prove stability in the L_∞-norm, i.e., to prove that bounded inputs produce bounded outputs. Mathematically, with $\|x\|_\infty = \sup_{t\in[0,\infty)} |x(t)|$, it is required to establish conditions under which the solutions of the feedback system (2.7) have the following property :

$$\|\sigma\|_\infty \leqslant \text{const.}\ \|f\|_\infty \ ; \ \|v\|_\infty \leqslant \text{const.}\ \|f\|_\infty .$$

Can one conclude that inequality (5.15) for the solutions of the system (2.7), established under the hypotheses of Theorem 5.31, implies L_∞-stability of (2.7) ? Compare your conclusions with those of Zames [Z 3(c)].

5.3. By suitable transformations or otherwise, derive (corresponding to the finite gain of the feedback part as in Exercises 2.1 and 2.2) L_2-stability conditions for the feedback system of Fig.5.1 in the following two separate forms : (i) the product, $\gamma(\mathcal{G}_1)\gamma(\mathcal{G}_2)$, of the gains of the operators $\mathcal{G}_1$ and $\mathcal{G}_2$, or the loop gain $\gamma(\mathcal{G}_1\mathcal{G}_2)$, of the composite operator $\mathcal{G}_1\mathcal{G}_2$ is less than 1 ; (ii) the operators $\mathcal{G}_1$ and $\mathcal{G}_2$ satisfy certain conic conditions (reminiscent of the sector conditions for scalar functions) of the form

$$\|(\mathcal{H}x)_T - \eta_1 x_T\| \gtreqless \eta_2 \|x_T\|$$

for all x in the domain of $\mathcal{H}$ and T in $[0,\infty)$ and for some constants $\eta_1 \geqslant 0$ and η_2. Relate your results to those of Zames [Z 3(d)] and of Sandberg [S 2(a) (e)].

5.4. Apply Theorem 5.31 to the feedback system (2.6), considered in Chapter 4, where now $f_o(t) \in L_2(0,\infty)$. Are L_2-stability and exponential stability of this system equivalent ? Also derive L_2-stability conditions (i) parallel to Theorem 4.42 for the feedback system (2.7) with $\varphi(\cdot) \in C$ and (ii) for the linear time invariant system (2.17) as a special case of Theorem 5.32. Compare the latter with the generalized Nyquist criterion (2.19)(Assumption S2, Chapter 2, p 31).

5.5. Suppose that k(t) of the feedback system (2.7) is not uniformly bounded over all time, i.e., approaches 0 for ∞ or oscillates with both as limit points as $t \to \infty$ [A 1]. Indicate how Theorem 5.31 can be generalized to apply to this case.

5.6. Given that the operator $\mathcal{N} \in \mathcal{O}$ (Definition 5.22) is anticipative. Prove (or disprove) that its adjoint $\mathcal{N}^*$ is nonanticipative. Comment on (or find conditions for) the existence of its inverse. Repeat the exercise for $\mathcal{N} \in \mathcal{O}$, nonanticipative and noncausal (Definition 3.21).

5.7. Suppose that the feedback system of Fig.5.1 is L_2-stable (Definition 2.38). Does it imply the existence of an operator $\mathcal{N}$ satisfying the hypotheses of Lemma 5.41 ? In other words, are the hypotheses of Lemma 5.41 necessary for the L_2-stability of the feedback system (5.17) ?

5.8. Consider the lightly damped Mathieu equation of Exercise 3.8 with a forcing function f(t) :

$$(dy^2/dt^2) + 2\mu(dy/dt) + (\mu^2+a)y + 16q(\cos 2t)y = f(t), \; t \geq 0. \qquad (5.21)$$

Suppose q and a are so chosen (based on the q-a stability chart of McLachlan [M 7] or Hayashi [H 6]) also to result in an L_2-stable system. Find the most general passive (possibly time varying) operator $\mathcal{N}$, satisfying the hypotheses of Lemma 5.41, where $\mathcal{G}_1$ represents the time invariant linear convolution operator with the Laplace transform $(1/(s+\mu)^2)$ and $\mathcal{G}_2$ the time varying gain, (a+16q cos 2t). Deduce the time invariant operator $Z_p \in \mathcal{O}$ defined by

$$(Z_p x)(t) = \sum_i (z_{ci} x(t-i\pi) + z_{aci} x(t+i\pi))$$

as a special case of the general (time varying) passive operator $\mathcal{N}$.

Further, with $\mathcal{N}$ constructed to meet the requirements specified above, can it be shown that, for q and a chosen in the unstable region of the q-a chart of [M 7] [H 6], $\mathcal{N}^* \mathcal{G}_2$ is not passive ? Hence or otherwise relate your conclusions to the results of Zames and Kallman [Z 5] who use the concept of 'spectral radius'.

5.9. For the system of Exercise 5.8, with q and a chosen to lie in the stability region of the q-a chart [M 7] [H 6], does there exist a time varying operator of the form $\alpha(t)+\beta(t)(d/dt)$ satisfying the hypotheses of Lemma 5.41 ?

Now, suppose that this form of the time varying operator is used to derive the relation between q and a for the L_2-stability of the lightly damped Mathieu equation (5.21). Compare this q-a relation with the stability chart of McLachlan [M 7] or Hayashi [H 6]. Hence comment on the results of Sundaresan and Thathachar [S 14(d)]. Also study the limiting case $\mu = 0$ for which the results of Zames and Kallman [Z 5] and others lead to trivial conclusions. Is it possible to infer that Lemma 5.41 (and hence Theorem 5.41) is not necessary for L_2-stability ?

5.10. For the feedback system (2.7), construct or give an example of a time varying passive operator satisfying the requirements of Lemma 5.41.

5.11. Let $L_p(0,\infty)$, where $p = 1,2,\cdots,\infty$ be the linear space of real valued functions $x(\cdot)$ on $[0,\infty)$ with the property that

$$\int_0^\infty |x(t)|^p \, dt < \infty \quad \text{if} \quad 1 \leq p < \infty$$

or $x(\cdot)$ is essentially bounded if $p = \infty$. Let $L_p(0,\infty)$ be equipped with the norm

$$\|x(\cdot)\|_p = \left(\int_0^\infty |x(t)|^p \, st\right)^{1/p} \quad \text{if} \quad 1 \leq p < \infty$$

and
$$\|x(\cdot)\|_\infty = \sup_{t\in[0,\infty)} |x(t)|.$$

For the feedback system (2.7), derive L_p-stability conditions which in the (limiting) case $p = 2$ reduce to Theorem 5.31. Compare with the results of Willems [W 3(e)] for the linear feedback system (2.23).

5.12. Consider the problem of analysing the stability of a system governed by the partial equation

$$(\partial y(t,r)/\partial t) = k(t,r)(\partial^2 y(t,r)/\partial r^2) + f(t,r), \tag{5.22}$$

where $r \in [0,1]$ is the independent (physical) space variable and $t \geq 0$ is time. Formulate this as an abstract Cauchy problem in a Banach space in which $y(t,r)$ is required to lie for each t, and write the equation as

$$(dy/dt) = A(t)y(t) + f(t) \tag{5.23}$$

where $A(t)$ is identified as the differential operator with the given boundary conditions (and is thus unbounded in general). How can one extend the results of (i) Chapter 3 to analyse the stability of (5.22) or (5.23) with zero input ; (ii) the present chapter to analyse the L_2-stability of (5.22) or (5.23) ? What is the relationship between the zero input stability and L_2-stability for this system?

5.13. An example given in [F 6] relates to the L_2-stability of a system governed by the partial differential equation

$$(\partial^4 y(t,r)/\partial t^2 \partial r^2) + 3(\partial^3 y/\partial t \partial r^2) + 2(\partial^2 y/\partial r^2) = N(y) + f(t,r) \tag{5.24}$$

where $N(\cdot)$ is the nonlinearity and $r \in [0,1]$ the space variable, and $t \geq 0$ time. Suggest suitable classes of input and output functions for which a stability definition can be given. Indicate how one can extend Theorem 5.41 and Lemma 5.41 to deal with the system (5.24). Note that Freedman et al [F 6] consider the special multiplier operator, E (unit operator), and hence their results are of limited application.

5.14. Consider the system governed by the partial differential equation

$$(\partial^2 y(t,r)/\partial t^2) + \alpha(\partial y/\partial t) - \beta(\partial^2 y/\partial r^2) = -(1+2\eta \cos r)N(y)$$

where α, β and η are constants, $r \in [0,1]$ the space variable, $t \geq 0$, the time and $N(\cdot)$ the nonlinearity. After a suitable choice of input and output functions, analyse the stability of the system and compare with the results of Cook [C 9] .

CHAPTER 6

INSTABILITY OF FEEDBACK SYSTEMS IN DIFFERENTIAL FORM

6.1. Introduction. After having had success of a sort with the stability analysis of the basic linear (and nonlinear) time varying feedback system, we now take up its instability analysis. Parallel to the contents of Chapter 3, we study the instability property of the linear feedback system

$$p(D)y + k(t)q(D)y = 0 \quad \text{on the interval } [0,\infty), \tag{2.21}$$

and of the nonlinear feedback system

$$p(D)y + k(t)\varphi(q(D)y) = 0 \quad \text{on the interval } [0,\infty), \tag{2.1}$$

in the Lyapunov-Chetaev-Corduneanu framework, under assumptions set forth in Chapter 2 (Sec.2.2 and Sec.2.3).

The main assumption we employ is Assumption U1. That is, we deal with Problem 2.42 of Chapter 2 (p 33). However, for completeness, a brief reference (Sec.6.2) is made to the question of instability under Assumption S1 also. It may be recalled that the assumptions needed in stability and instability analyses refer to the properties of the linear time invariant system

$$p(D)y + Kq(D)y = 0$$

or, equivalently, (2.15)

$$(d\underline{x}/dt) = A_o\underline{x} - K\,\underline{b}\,\underline{c}'\,\underline{x}$$

where the constant K is assumed, for convenience in manipulations, to take values in $[0,\infty)$. Any other range of variation for K can be converted to this 'infinite sector' by a suitable transformation (Exercise 2.2).

Analogous to the stability conditions of Chapter 3, we seek to express instability conditions for systems (2.1) and (2.21) in terms of the frequency characteristic $G(j\omega)$ and a suitable constraint on $k(t)$. To this end, we choose a quadratic form for the linear system (2.21), and a quadratic form + a certain integral of the nonlinearity for the nonlinear system (2.1), and then evaluate their time-derivatives along the solutions of (2.21) and (2.1) respectively. The time-derivatives so evaluated are rewritten in the form of a suitable differential inequality. We then invoke the Lyapunov-Chetaev (Lemma

6.31) or the Lyapunov-Corduneanu (Lemma 6.32) instability lemma depending on the property of the quadratic form (+ integral of the nonlinearity) chosen earlier. The quadratic forms are generated much as in Sec.3.3 of Chapter 3.

It turns out that an application of the Lyapunov-Corduneanu instability lemma holds the promise of more general instability conditions which are counterparts of the stability conditions of Chapter 3. However, the contents of the present chapter (as also of the next) are better treated by the reader as tentative material, needing further clarification and confirmation. A reason for this is the conflicting nature of results available in the literature.

With the above proviso, Sec.6.4 and Sec.6.5 contain instability results obtainable respectively from the Lyapunov-Chetaev and the Lyapunov-Corduneanu instability lemmas. A comparison is made in Sec.6.6 with the results available in the literature. The conclusion (Sec.6.7) is that considerable basic work remains to be done in the instability analysis of the system (2.1).

6.2. Possible Approaches to Instability Analysis.

The linear time invariant system (2.15) with K in $[0,\infty)$ may have the following types of behaviour of its solutions :

Case (a). $\| \underline{x}(t; t_o, \underline{x}_o) \| \leq \eta_1 \| \underline{x}_o \| \exp(-\eta_2(t-t_o))$ (6.1)

Case (b). $\| \underline{x}(t; t_o, \underline{x}_o) \| \leq \eta_1 \| \underline{x}_o \|$ (6.2)

Case (c). $\| \underline{x}(t; t_o, x_o) \| \geq \eta_1 \| \underline{x}_o \| \exp(\eta_2(t-t_o))$ (6.3)

for some positive constants η_1 and η_2, and for all $t \geq t_o \geq 0$. Correspondingly, the eigenvalues of the system matrix $A_1 \triangleq (A_o - K \underline{b} \underline{c}')$ with K in $[0,\infty)$ have the following properties :

Case (a). All the eigenvalues of A have negative real parts.

Case (b). Some eigenvalues of A have negative real parts and others have zero real parts.

Case (c). At least one eigenvalue of A has a positive real part.

In practice, the system (2.15) has the solution property (6.1) for some finite K in $[0,\infty)$ and the solution property (6.3) for other values of K in in $[0,\infty)$. However, for simplicity and convenience in manipulations, we assume that the system (2.15) has been suitably transformed (Exercise 2.2) so that the solutions have one of the properties (6.1)-(6.3) for all K in $[0,\infty)$.

According to the fundamental theorem of Lyapunov [H 3(a), pp 33-34] on instability in the 'first approximation', if the 'Aizerman system' (2.15) has the solution behaviour (6.3) for all K in $[0,\infty)$, then the nonlinear time varying system (2.1) or, in the vector form,

$$(d\underline{x}/dt) = A_o\underline{x} - k(t)\,\underline{b}\,\varphi(\underline{c}'\underline{x}) \tag{2.4}$$

also has the solution behaviour given by (6.3) provided

$$|k(t)\varphi(\sigma)| < h\,|\sigma| \tag{6.4}$$

for some sufficiently small constant h, for all $\sigma \neq 0$ and $t \geq 0$. Such an upper bound on $k(t)\varphi(\cdot)$ is of little use in practice. If the matrix A has some eigenvalues with zero real parts and others with positive real parts, an interesting question is the following : Is the system (2.4) unstable for arbitrary $k(t)\varphi(\cdot)$ satisfying the inequality

$$0 < \varepsilon \leq (k(t)\varphi(\sigma)/\sigma) < \infty \tag{2.20}$$

for some $\varepsilon > 0$ and all $t \geq 0$ and $\sigma \neq 0$?

A natural approach is to convert the instability problem to a stability problem so that results of Chapter 3 could be used. The methods based on this idea are listed in Table 6.1.

Table 6.1

No.	Original system being analysed for instability (OSI)	Equivalent system to be analysed for stability (ESS)	Relationship between the solutions of OSI and ESS	Comments
1.	Linear system (2.21).	Adjoint system $(d\underline{\xi}/dt) = -A_o'\underline{\xi} + k(t)\underline{c}'\underline{b}\,\underline{\xi}$	$\underline{x}'(t)\underline{\xi}(t) = \underline{x}'(t_o)\underline{\xi}(t_o)$. If the components of $\underline{\xi}(t)$ are decreasing, those of $\underline{x}(t)$ are increasing and vice	When some of the components of $\underline{\xi}(t)$ are decreasing and others increasing, the components of $\underline{x}(t)$ are increasing and decreasing correspondingly. Further, the stability conditions for ESS may not be available (Exercise 6.1). Even when they are available, the resulting

No.	Original system being analysed for instability (OSI)	Equivalent system to be analysed for stability (ESS)	Relationship between the solutions of OSI and ESS	Comments
			versa.	instability conditions may be overly conservative (Exercise 6.2).
2.	Nonlinear system (2.1).	Adjoint system not defined.	-----	-----
3.	Linear system (2.21).	System obtained by reversing the sign of t in (2.21) : $(d\underline{\zeta}/dt) = -A_o\underline{\zeta} + k(-t)\underline{b}\,\underline{c}'\underline{\zeta}$	$\underline{x}(t) = \underline{\zeta}(-t)$	See item No.1 above. (Exercise 6.2).
4.	Nonlinear system (2.21).	System obtained by reversing the sign of t in (2.1): $(d\underline{\zeta}/dt) = -A_o\underline{\zeta} + k(-t)\underline{b}\,\varphi(\underline{c}'\underline{\zeta})$.	$\underline{x}(t) = \underline{\zeta}(-t)$	See item No.1 above.

A classical result which could occasionally be used in the instability analysis of the linear system (2.21) is the one concerning the Wronskian [K 4, Chapter 8, pp 436-439] . See also Exercise 3.16, p 89. Suppose in the system (2.21), the order m of the polynomial q(s) is equal to n-1, where n is the order of the polynomial p(s). Then, the equation (2.21) could be written as

$$(d^n y/dt^n) + (p_{n-1} + k(t)q_{n-1})(d^{n-1}y/dt^{n-1}) + (p_{n-2} + k(t)q_{n-2})$$
$$(d^{n-2}y/dt^{n-2}) + \cdots + (p_o + k(t)q_o)y = 0 \quad \text{on } [0,\infty), \tag{6.5}$$

and the corresponding Wronskian W(t)(= determinant of the system transition matrix) satisfies the relation

$$W(t) = W(t_o)\exp(-\int_{t_o}^{t} (p_{n-1} + k(\tau)q_{n-1})d\tau) \tag{6.6}$$

Hence the system (6.5) is exponentially unstable (Exercise 6.3) if

$$\int_{t_o}^{t} (p_{n-1} + k(\tau)q_{n-1})d\tau \leq -\varepsilon(t-t_o) \tag{6.7}$$

for some positive constant ε and for all $t \geqslant t_o$. Note that the nature of the other coefficients $p_{n-2}, p_{n-3}, \ldots, p_o, q_{n-2}, q_{n-3}, \ldots, q_o$ is immaterial to the instability behaviour of the system (6.5).

In case $m < (n-1)$, the instability criterion (6.7) is to be modified to the requirement that $p_{n-1} < 0$. For $p_{n-1} > 0$ and $m < (n-1)$, the above method fails (Exercise 6.3). Note that no similar result holds for the nonlinear system (2.1). The inference is that a more general approach should be tried to resolve the problem of instability.

6.21. The derivation of instability criteria has been justified by some authors [B 18] [N 3(b)(c)] as follows : The stability criteria (for instance, those of Chapter 3) are sufficient but not necessary. As it is not known how to establish necessity of these criteria straightaway, the proposed instability criteria indicate the extent to which the stability criteria are necessary as well.

A closer inspection of the literature reveals that almost all of the instability criteria have been derived under assumptions differing from those used in stability analysis. This is clarified in Table 6.2 which shows the assumptions ptions made and the corresponding stability and instability criteria for the system (2.21) which are derivable or already derived or still unknown. Evidently, the contents of Table 6.2 hold, mutatis mutandis, for the nonlinear system (2.1).

The next section introduces a general approach to instability analysis.

Table 6.2

No.	System under consideration	Basic assumptions made about the 'Aizerman system' (2.15)	Nature of derivable or already derived (or desired, but unknown) criteria.	Basic lemma used for deriving the criteria. Comments, if any.
1.	Linear system (2.21)	Solutions with property (6.1) for all K in $[0,\infty)$. (Assumption S1, Sec. 2.32, p 30)	Exponential stability (solution to Problem 2.41, Sec. 2.4, p 32). Also asymptotic stability criteria of literature referred to in Chapter 3.	Lyapunov-Corduneanu (Lemma 3.33, p 64). Lyapunov.

No.	System under consideration	Basic assumption made about the 'Aizerman system'(2.15)	Nature of derivable or already derived (or desired, but unknown) criteria.	Basic lemma used for deriving the criteria. Comments, if any.
2	Linear system (2.21)	Solutions with property (6.1) for all K in [0,∞)(Assumption S1, Sec.2.32 2.32, p 30).	Instability criteria (solution to Problem 2.44, Sec.2.4, p 34). Still an unsolved problem	The problem is nontrivial (Exercises 6.4 and 6.16) See also Chapter 1, Sec.1.21(c).
3	Linear system (2.21)	Solutions with property (6.3) for all K in [0,∞)(Assumption U1, Sec. 2.32, p 31).	Exponential instability criteria, attempted in the present chapter (Sec. 6.3 and Sec.6.4). Also most of the instability criteria in the literature.	Lyapunov-Chetaev (Lemma 6.31) and Lyapunov-Corduneanu (Lemma 6.32).
4	Linear system (2.21)	Solutions with property (6.3) for all K in [0,∞).(Assumption U1, Sec. 2.32, p 31).	Stability criteria (solution to problem 2.43, Sec.2.4, p 34). Still an unsolved problem.	The problem is nontrivial (Exercise 6.5).

6.3. A General Approach to Instability Analysis.

One of the early results on instability of the linear system (2.21) or, equivalently,

$$(d\underline{x}/dt) = A_o\underline{x} - k(t)\,\underline{b}\,\underline{c}'\underline{x} \qquad (2.21)$$

is usually credited to Wazewski [Z 2, Chapter 7, pp 379-382]. The main idea of the result is this:

Let $v(\underline{x}) = \underline{x}'P\,\underline{x}$ where P is a constant positive definite matrix. Its time-derivative along the solutions of (2.21) is given by

$$(dv(\underline{x})/dt)\Big|_{(2.21)} = \underline{x}'(t)(A_o'P + P\,A_o)\,\underline{x}(t) - k(t)\underline{x}'(t)(\underline{c}\,\underline{b}'P + P\,\underline{b}\,\underline{c}')\underline{x}(t)$$

which satisfies the inequality

$$(\lambda_{1\,min} - k(t)\,\lambda_{2\,max})v(\underline{x}(t)) \leq (dv(\underline{x})/dt)\Big|_{(2.21)} \leq (\lambda_{1\,max} - k(t)\lambda_{2min})v(\underline{x}(t))$$

for all $t \geq 0$, where

$$\lambda_{1min} = \min_{\underline{x}} (\underline{x}'(A_o'P + P\,A_o)\underline{x}/\underline{x}'P\,\underline{x}) ;$$

$$\lambda_{1\,max} = \max_{\underline{x}} (\underline{x}'(A_o'P + P\,A_o)\underline{x}/\underline{x}'P\,\underline{x}) ;$$

$$\lambda_{2min} = \min_{\underline{x}} (\underline{x}'(\underline{c}\,\underline{b}'P + P\,\underline{b}\,\underline{c}')\underline{x}/\underline{x}'P\,\underline{x}) ; \text{ and}$$

$$\lambda_{2max} = \max_{\underline{x}} (\underline{x}'(\underline{c}\,\underline{b}'P + P\,\underline{b}\,\underline{c}')\underline{x}/\underline{x}'P\,\underline{x}).$$

If $(\lambda_{1\,min} - k(t)\lambda_{2\,max})$ satisfies the integral inequality

$$\int_{t_o}^{t} (\lambda_{1\,min} - k(\tau)\,\lambda_{2\,max})\, d\tau \geq \varepsilon(t-t_o) \qquad (6.8)$$

for some constant $\varepsilon > 0$ and for all $t \geq t_o \geq 0$, then the system (2.21) is exponentially unstable. If $\varepsilon = 0$ in the integral inequality (6.8), then the system (2.21) is merely unstable in the Lyapunov sense (Definition 2.36, p 27).

In spite of the apparent generality of the above instability criterion, it becomes evident, on application to a specific problem, that the instability condition so obtained is very conservative (Exercise 6.6) even for the linear time invariant system (2.15). An extension of the Wazevski idea to the nonlinear time varying system (2.1) suffers from the same weakness.

6.31. Lyapunov-Chetaev and Lyapunov-Corduneanu Instability Lemmas.

The stage is set for introducing a general approach to instability analysis. The main tools are the following two lemmas which lead to somewhat different types of instability criteria for the systems (2.1) and (2.21).

Lemma 6.31 (Lyapunov-Chataev) [H 3(a), pp 18-19].

Given the differential equation (2.1) (or, equivalently, (2.4)), and a real valued function $v_1(\underline{x},t)$ with the following properties.

(a) There exist at each time t points arbitrarily close to $\underline{x} = \underline{0}$ at which $v(\underline{x},t) < 0$ for all $t \geq t_o \geq 0$, the totality of points $(\underline{x},t)$ with $\|\underline{x}\| < h$ and $v_1(\underline{x},t) < 0$ being denoted by the domain $\mathcal{D}$.

(b) In a sub-domain $\mathcal{D}_1$ of $\mathcal{D}$,

$$v_1(\underline{x},t) \geq -\eta_1 \|\underline{x}\|^2 \tag{6.9}$$

for some constant $\eta_1 > 0$.

(c) In the sub-domain $\mathcal{D}_1$ of trajectories, the time-derivative of $v_1(\underline{x},t)$ with respect to (2.1) (or, equivalently, (2.4)) satisfies the inequality

$$(dv_1(\underline{x},t)/dt)\Big|_{(2.1)} \leq -\eta_2 \|\underline{x}(t)\|^2 \tag{6.10}$$

for some constant $\eta_2 > 0$.

Then the system (2.1) has solutions with the property that

$$\|\underline{x}(t)\|^2 \geq (\eta_3 \|\underline{x}_o\|^2 + \eta_4 \int_{t_o}^{t} \|\underline{x}(\tau)\|^2 \, d\tau) \tag{6.11}$$

for some positive constants η_3 and η_4, and for all $t \geq t_o \geq 0$ until the boundary $\|\underline{x}\| = h$ of the domain $\mathcal{D}$ is reached.

Proof. Integration of the differential inequality (6.10) yields

$$v_1(\underline{x}(t),t) \leq v_1(\underline{x}_o,t_o) - \eta_2 \int_{t_o}^{t} \|\underline{x}(\tau)\|^2 \, d\tau \tag{6.12}$$

for all $t \geq t_o \geq 0$. Choosing the initial point $(\underline{x}_o, t_o)$ in the domain $\mathcal{D}_1$, we get from (6.12) the inequality

$$v_1(\underline{x}(t),t) \leq -\eta_o \|\underline{x}_o\|^2 - \eta_2 \int_{t_o}^{t} \|\underline{x}(\tau)\|^2 \, d\tau \tag{6.13}$$

for some positive constant $\eta_o < \eta_1$, for all $t \geq t_o \geq 0$ until the boundary $\|\underline{x}\| = h$ is reached. By virtue of the inequality (6.9), inequality (6.13) becomes

$$-\eta_1 \|\underline{x}(t)\|^2 \leq -(\eta_o \|x_o\|^2 + \eta_2 \int_{t_o}^{t} \|\underline{x}(\tau)\|^2 \, d\tau) \tag{6.14}$$

for all $t \geq t_o \geq 0$ until the boundary $\|\underline{x}\| = h$ is reached. Inequality (6.14) can be rewritten in the form of inequality (6.11). The lemma is proved.

Corollary 6.311. Suppose in place of the differential inequality (6.10), we have

$$(dv_1(\underline{x},t)/dt)\Big|_{(2.1)} \leq -\eta_2\Big(\sum_i \gamma_i x_i(t)\Big)^2 \tag{6.15}$$

for some constants $\eta_2 > 0$ and γ_i, $i = 1,2,\cdots m$ where $m \leq n$. Then the solution inequality (6.11) becomes

$$\|\underline{x}(t)\|^2 \geq \eta_3 \|\underline{x}_o\|^2 + \eta_4 \int_{t_o}^{t} \Big(\sum_i \gamma_i\, x_i(\tau)\Big)^2 d\tau \tag{6.16}$$

for some positive constants η_3 and η_4, for all $t \geq t_o \geq 0$ until the boundary $\|\underline{x}\| = h$ is reached.

Lemma 6.32. (Lyapunov-Corduneanu) [C 8] .

If there exist a quadratic form $v_2(\underline{x},t) = \underline{x}'P(t)\underline{x}$ having the property

$$\gamma_o \underline{x}'\underline{x} \leq v_2(\underline{x},t) \leq \gamma_1 \underline{x}'\underline{x} \tag{6.17}$$

for some positive constants γ_o and γ_1 with $\gamma_o < \gamma_1$, for all $t \geq t_o \geq 0$, and a real valued function $\lambda(t)$ on $[t_o,\infty)$ such that the time-derivative of $v_2(\underline{x},t)$ along the solutions of (2.1) (or, equivalently, of (2.4)) satisfies the inequality

$$(dv_2(\underline{x},t))\Big|_{(2.1)} \geq \lambda(t)v_2(\underline{x}(t),t) \tag{6.18}$$

then there exists a positive constant η_o such that the solutions of (2.1) (and hence, of (2.4)) satisfy the inequality

$$\|\underline{x}(t)\| \geq \eta_o \|\underline{x}(t_o)\| \exp\Big(\frac{1}{2}\int_{t_o}^{t} \lambda(\tau)d\tau\Big) \tag{6.19}$$

for all $t \geq t_o \geq 0$.

Proof. Similar to the proof of Lemma 3.33. See Appendix 6.1.

Corollary 6.321. If, in Lemma 6.32, $\lambda(\cdot)$ satisfies the inequality

$$(1/T)\int_{t_o}^{t_o+T} \lambda(\tau)d\tau \geq \varepsilon \tag{6.20}$$

for some positive constant ε and for all $T > 0$, then

$$\|\underline{x}(t)\| \geq \eta_o \|\underline{x}_o\| \exp(\varepsilon(t-t_o)/2)$$

which implies that the system (2.1) is exponentially unstable.

Corollary 6.322. Suppose, in Lemma 6.32, (i) $v_2(\underline{x},t)$ is nonquadratic and has the property

$$\gamma_o \underline{x}'\underline{x} \leqslant v_2(\underline{x},t) \leqslant \gamma_1 \underline{x}'\underline{x} + \gamma_2 \exp(\alpha t) \int_{t_o}^{t} \exp(-\alpha\tau) \| \underline{x}(\tau) \|^2 d\tau \qquad (6.21)$$

for some positive constants γ_o and γ_1 with $\gamma_o < \gamma_1$, and nonnegative constants γ_2 and α , for all $t \geqslant t_o \geqslant 0$; and (ii) $\lambda(\cdot)$ satisfies the integral inequality

$$(1/T) \int_{t_o}^{t_o+T} \lambda(\tau) d\tau \geqslant \alpha - \varepsilon$$

for some $\varepsilon > 0$, for all $T > 0$. Then, the solutions of (2.1) (and, hence, of (2.4)) have the property

$$\int_{t_o}^{t} \exp(-\alpha\tau) \| \underline{x}(\tau) \|^2 d\tau \geqslant \gamma_3 \|x_o\|^2 (\exp(-\varepsilon(t-t_o)) - \exp(-\gamma_2(t-t_o)/\gamma_1) \qquad (6.22))$$

for some positive constant γ_3, for all $t \geqslant t_o \geqslant 0$.

Proof. See Appendix 6.2.

Remark 6.31. Lemmas 6.31 and 6.32 will be used below (along with the method of generating quadratic forms of Sec. 3.3 and Sec. 3.4) to derive instability criteria for the system (2.1). It turns out that the former enables us to derive instability criteria in multiplier form, the multiplier function being either 1 or a + bs, with the restriction that an arbitrary (positive real) multiplier function is inadmissible. In contrast, Lemma 6.32 does permit derivation of instability criteria for the system (2.1) in terms of an arbitrary positive real multiplier function. A comparison between the two types of instability criteria will be made later (Table 6.3). Note that Lemma 6.31 was first used by Brockett and Lee [B 18] .

6.4. Derivation of Instability Criteria

Part 1. Application of the Lyapunov-Chetaev Lemma.

Guided by the stability results of Chapter 3, we imagine that the derivation of general instability criteria requires an enlargement of the state space of the system (2.1). That is,we have to consider an auxiliary system in whose state space the state space of (2.1) is imbedded. As in the case of stability analysis, the auxiliary system should be such that no extraneous instability has been introduced affecting the original system behaviour.

In what follows, $Z(s) = m(s)/n(s)$ where $m(s)$ and $n(s)$ are finite polynomials in s.

6.41. Generation of Quadratic Forms.

Based on the results of Sec.3.3, we generate quadratic forms assuming that $\text{Re}\, Z(j\omega-\alpha) \geq 0$ and $\text{Re}\, Z(j\omega-\beta)\, G(j\omega-\beta) \geq 0$ for some constants α and β, for all real ω.

Let $\underline{\chi}(t)$ denote the state vector of the system

$$p(D)n(D)y + k(t)\varphi(q(D)n(D)y) = 0 \,. \tag{6.23}$$

The state vector $\underline{x}(t)$ of (2.1) is a subspace of $\underline{\chi}(t)$. Let

$$r_1(s) = (\text{Ev}\; q(s-\beta)m(s-\beta)p(-s-\beta)n(-s-\beta))^{(-)},\; r_2(s) = (\text{Ev}\; m(s-\alpha)n(-s-\alpha))^{(-)}.$$

A suitable choice of the integration path (see Sec.3.4) leads to the following quadratic forms in $\underline{\chi}$:

$$V_1(\underline{\chi},t) = \exp(-2\beta t)\int_{t(\underline{0})}^{t(\underline{\chi})}\Big\{p(D-\beta)n(D-\beta)(\chi\exp(\beta\tau))q(D-\beta)m(D-\beta)(\chi\exp(\beta\tau)) - (r_1(D)(\chi\exp(\beta\tau))^2\Big\}\, d\tau \,, \tag{6.24}$$

$$V_2(\underline{\chi},t) = \exp(-2\alpha t)\int_{t(\underline{0})}^{t(\underline{\chi})}\Big\{m(D-\alpha)(\exp(\alpha\tau)q(D)\chi)n(D-\alpha)(\exp(\alpha\tau)q(D)\chi) - (r_2(D)(\exp(\alpha\tau)q(D)\chi))^2\Big\}\, d\tau \,. \tag{6.25}$$

An extension of Lemma 3.31 and Corollary 3.311 leads to the result (Exercise 6.7) that if (i) $Z(s-\alpha)$ is not positive real but $\text{Re}\, Z(j\omega-\alpha) \geq 0$ for some constant α, for all real ω, $V_2(\underline{\chi},t)$ is sign indefinite ; and (ii) $Z(s-\beta)G(s-\beta)$ is not positive real but $\text{Re}\, Z(j\omega-\beta)G(j\omega-\beta) \geq 0$ for some constant β, for all real ω, $V_1(\underline{\chi},t)$ is also sign indefinite.

The quadratic form $V_1(\underline{\chi},t) + k(t)V_2(\underline{\chi},t)$ is used for the instability analysis of the linear system (2.21). For the instability analysis of the nonlinear system, we must have recourse to the nonquadratic form $V_1(\underline{\chi},t) + k(t)$ (integral involving the nonlinearity $\varphi(\cdot)$). The latter will be introduced below.

6.42. Derivation of Instability Criteria.

The relationship between the instability property of the auxiliary system (6.23) and that of the original system (2.1) is given by the following lemma.

Lemma 6.41. If the system (6.23) has an exponentially unstable solution, and $n(D)w = 0$ represents an asymptotically stable system, then the system (2.1) also has an exponentially unstable solution.

Proof. If $\phi(t)$ is a solution of (6.23), then $n(D)\phi(t)$ is a solution of (2.1). If $\phi(t)$ increases exponentially, so do its first $\eta + n - 1$ derivatives. Denote $n(D)\phi(t)$ by $\phi'(t)$. We conclude that $\phi'(t)$ and its first (n-1) derivatives increase exponentially. The lemma is proved.

Corollary 6.411. If the linear system

$$p(D)n(D)y + k(t)q(D)n(D)y = 0 \qquad (6.26)$$

has an exponentially unstable solution, and $n(D)w = 0$ represents an asymptotically stable system, then the system (2.21) has an exponentially unstable solution.

Remark 6.41. Suppose $n(D)w = 0$ represents an unstable system. Then the original system (2.1) could be stable, while the auxiliary system (6.23) unstable. Hence the conclusion of Lemma 6.41 need not hold in this case.

We now consider the derivation of instability criteria under two separate basic assumptions about the linear time invariant system (2.15). See Table 6.2 for details.

Case 1. Instability Criteria based on Assumption S1.

Assumption S1 along with the multiplier form of the Nyquist criterion (Lemma 3.22) implies that the quadratic forms $V_1(\underline{\chi}, t)$ and $V_2(\underline{\chi}, t)$ defined by (6.24) and (6.25) are positive definite and positive semidefinite respectively for some positive constants α and β. But we need sign indefinite quadratic forms for invoking the Lyapunov-Chetaev instability lemma. Hence Assumption S1 is untenable.

Case 2. Instability Criteria based on Assumption U1.

Assumption U1 along with the multiplier form of the Nyquist criterion implies that $\arg\{Z(j\omega)\}$ lies beyond the band $[-\pi/2, \pi/2]$ in nontrivial frequency intervals. Hence the inequality $\mathrm{Re}\, Z(j\omega) \geq 0$ for all real ω is violated. But an attempt to use the inequality $\mathrm{Re}\, Z(j\omega + \alpha) \geq 0$ for some $\alpha > 0$, for all real ω will lead to such changes in the auxiliary system (6.23) as to imply the use of positive realness conditions for generation of quadratic forms (Exercise 6.8). Hence the Lyapunov-Chetaev lemma, which is based on sign indefinite quadratic forms, cannot be invoked. This accounts for the lack of success in the extension of the Brockett and Lee instability criterion (derived below), where by the term 'extension' we mean the use of an arbitrary multiplier function in the statement of instability criteria.

There are two special multiplier functions which can be employed under a suitable behaviour of $\arg\{G(j\omega)\}$:

(i) Suppose $\text{Re}\ G(j\omega) \gtrsim 0$ for all real ω but $G(s)$ is not positive real. Then 1 is an instability multiplier.

(ii) Suppose $\text{Re}(a_o + b_o j\omega)G(j\omega) \gtrsim 0$ for all real ω , but $(a_o + b_o s)G(s)$ is not positive real. Then $(a_o + b_o s)$ is an instability multiplier.

In both these instances, $n(s) = 1$ and hence the auxiliary system (6.23) is simply the original system (2.1).

We now establish two instability criteria.

Theorem 6.41. The system (2.1) has solutions with property (6.16) if (i) $\text{Re}\ G(j\omega) \gtrsim 0$ for all real ω and (ii) $G(s)$ is not positive real.

Proof. The quadratic form

$$V_{cl}(\underline{x}) = \int_{t(\underline{0})}^{t(\underline{x})} \left\{ p(D)x\ q(D)x - (r_{cl}(D)x)^2 \right\} d\tau$$

where $r_{cl}(s) = (\text{Ev}\ q(s)p(-s))^{(-)}$, is sign indefinite by virtue of hypothesis (ii). Its time-derivative along the solutions of (2.1) is given by

$$(dV_{cl}(\underline{x})/dt) \Big|_{(2.1)} = -k(t)\varphi(q(D)y(t))(q(D)y(t)) - (r_{cl}(D)y(t))^2$$

Invoke Lemma 6.31 and its Corollary 6.311 to conclude that the system (2.1) has solutions with property (6.16). The theorem is proved.

Remark 6.42. The finite gain version of Theorem 6.41 is the circle instability criterion of Brockett and Lee [B 18] . The requirement that $\text{Re}\ G(j\omega) \gtrsim 0$ for all real ω is too stringent. This is weakened to some extent in the next theorem.

Theorem 6.42. The system (2.1) with $\varphi(\cdot) \in C$ has solutions with property (6.16) if

(i) $\mathrm{Re}(a_o + b_o j\omega)G(j\omega) \geq 0$ for some constants $a_o > 0$ and $b_o \geq 0$, for all real ω;

(ii) $(a_o + b_o s)G(s)$ is not positive real for values of a_o and b_o satisfying hypothesis (i) ; and

(iii) $(dk/dt) \leq \alpha_o k(t)$ for some positive constant $\alpha_o \leq (a_o/b_o\delta_s)$.

Proof. The quadratic form

$$V_{c2}(\underline{x}) = \int_{t(\underline{0})}^{t(\underline{x})} \left\{ p(D)x\ q(D)(a_o + b_o D)x - (r_{c2}(D)x)^2 \right\} d\tau \tag{6.27}$$

where $r_{c2}(s) = (\mathrm{Ev}\ q(s)(a_o + b_o s)p(-s))^{(-)}$, is sign indefinite by virtue of hypothesis (ii). On the basis of Lemma 3.51, the functional

$$V_{c3}(\underline{x},t) = \exp(-\alpha_o t)\int_0^t \exp(\alpha_o \tau)\ \varphi(q(D)x)((a_o + b_o D)q(D)x)\ d\tau \tag{6.28}$$

is nonnegative for $a_o > 0$, $b_o \geq 0$ if $0 < \alpha_o \leq a_o/b_o\delta_s$.

Define

$$V_c(\underline{x},t) = V_{c2}(\underline{x}) + k(t)V_{c3}(\underline{x},t) \tag{6.29}$$

where $V_{c2}(\underline{x})$ and $V_{c3}(\underline{x},t)$ are defined by (6.27) and (6.28) respectively. It is easy to verify that $V_c(\underline{x},t)$ satisfies hypotheses (a) and (b) of Lemma 6.31 in view of the fact $V_{c2}(\underline{x})$ is sign indefinite and $\lim_{t\to\infty} V_{c3}(\underline{x},t) = \varphi(q(D)x)(a_o + b_o D)q(D)x/\alpha_o$. The time-derivative of $V_c(\underline{x},t)$ along the solutions of (2.1) becomes

$$(dV_c(\underline{x},t)/dt)\Big|_{(2.1)} = ((dk/dt) - \alpha_o k(t))\ V_{c3}(\underline{x},t) - (r_{c2}(D)\ x)^2$$

which, by virtue of hypothesis (iii) satisfies an inequality of the type (6.15) of Corollary 6.311. Now invoking Lemma 6.31 and its Corollary 6.311, we conclude that the system has solutions with property (6.16). The theorem is proved.

Remark 6.43. A geometric interpretation of hypothesis (i) of Theorem 6.42 is the same as the well known Popov criterion (see Sec. 3.5, p 79). An additional requirement is that $(a_o + b_o s)G(s)$ should not be positive real. The point-by-point constraint on (dk/dt) of hypothesis (iii) can be transformed to a global constraint by the technique used in Chapter 3 (Lemma 3.42, p 69). Note that the Nyquist instability criterion for the linear time invariant system (2.15) is not obtainable as a special case of Theorems 6.41 and 6.42, Further, the instability counterpart of Theorem 3.41 (pp 67-68) is also not available.

6.5. Derivation of Instability Criteria.

Part II. Application of the Lyapunov-Corduneanu Lemma.

The main idea in the application of Lemma 6.32 is the following basic result :

If Assumption U1 (Sec.2.32, pp 30-31) holds, then the damped auxiliary linear time-invariant system

$$p(D+\beta)u + K\,q(D+\beta)u = 0 \quad \text{on } [0,\infty) \tag{6.30}$$

where K is a constant in $[\varepsilon,\infty)$ for some $\varepsilon > 0$, is not stable for any non-positive value of β, but is asymptotically stable for some positive $\beta \geqslant \beta_i > 0$. Here β_i is the minimum value of β for which the system (6.30) is asymptotically stable. If there is no finite value of β for the asymptotic stability of (6.30), we set $\beta_i = \infty$.

For the system (6.30) with $\beta \geqslant \beta_i$, we can construct a strictly positive real multiplier function satisfying the conditions of Lemma 3.22 (p 57).

Before stating the main results of the section, we need to settle some preliminaries.

Lemma 6.51. If the nonlinear system

$$p(D)n(D-\beta)y + k(t)\varphi(q(D)n(D-\beta)y) = 0 \quad \text{on } [0,\infty) \tag{6.31}$$

is exponentially unstable for some $\beta > 0$, and $n(0) \neq 0$, then the system (2.1) is also exponentially unstable.

Proof. See Appendix.6.3.

Corollary 6.511. If the linear system

$$p(D)n(D-\beta)y + k(t)q(D)n(D-\beta)y = 0 \quad \text{on } [0.\infty) \tag{6.32}$$

is exponentially unstable for some $\beta > 0$, and $n(0) \neq 0$, then the system (2.21) is also exponentially unstable.

We deal first with the instability analysis of the linear system (2.21) and next with that of the nonlinear system (2.1).

6.51. Linear System Instability Analysis.

For the generation of suitable positive definite quadratic forms required by Lemma 6.32, we assume that (i) $Z(s)G(s+\beta)$ is positive real for some $\beta > 0$, and $q(s+\beta)m(s)$ (or $p(s+\beta)n(s)$) and $r'(s) = (\text{Ev } q(s+\beta)\ m(s)p(-s+\beta)n(-s))^{(-)}$ do not have common factors on the imaginary axis; (ii) $Z(s+\alpha-\beta)$ is strictly positive real for some $\alpha \geq \beta$. Let $r''(s) = (\text{Ev } m(s+\alpha-\beta)n(-s+\alpha-\beta))^{(-)}$. The path of integration needed in the generation of a quadratic form is assumed to be suitably chosen (Sec.3.3). Then based on the results of Sec.3.3, we have the following positive definite quadratic forms in $\underline{\chi}$:

$$V'(\underline{\chi},t) = \exp(2\beta t)\int_{t(\underline{0})}^{t(\underline{\chi})}\Big\{q(D+\beta)m(D)(\chi\exp(-\beta\tau))p(D+\beta)n(D)(\chi\exp(-\beta\tau)) - (r'(D)(\chi\exp(-\beta\tau))^2\Big\}d\tau, \tag{6.33}$$

$$V''(\underline{\chi},t) = \exp(2\alpha t)\int_{t(\underline{0})}^{t(\underline{\chi})}\Big\{m(D-\beta+\alpha)q(D+\alpha)(\chi\exp(-\alpha\tau))n(D-\beta+\alpha)q(D+\alpha)(\chi\exp(-\alpha\tau)) - (r''(D)q(D+\alpha)(\chi\exp(-\alpha\tau))^2\Big\}d\tau \tag{6.34}$$

Let, as before, $\theta(t) \triangleq ((dk/dt)/k(t))$. The time-derivative of the quadratic form

$$V(\underline{\chi},t) = V'(\underline{\chi},t) + k(t)\,V''(\underline{\chi},t) \tag{6.35}$$

along the solutions of (6.32) is given by

$$(dV(\underline{\chi},t)/dt)\Big|_{(6.32)} = 2\beta\,V'(\underline{\chi}(t),t) + (\theta(t) + 2\alpha)k(t)V''(\underline{\chi}(t),t) - (r'(D-\beta)y(t))^2 - k(t)(r''(D-\alpha)q(D)y(t))^2 \tag{6.36}$$

which satisfies the inequality

$$(dV(\underline{\chi},t)/dt)\Big|_{(6.32)} \geq \inf_{t\geq 0}\,(2\beta,\ \theta(t) + 2\alpha)\,V(\underline{\chi}(t),t) - \max(\eta_1,\ \eta_2)\,V(\underline{\chi}(t),t) \tag{6.37}$$

where

$$\eta_1 = \max_{\underline{\chi}}((r'(D-\beta)\chi)^2/V'(\underline{\chi},t)),$$
$$\eta_2 = \max_{\underline{\chi}}((r''(D-\alpha)q(D)\chi)^2/V''(\underline{\chi},t)). \tag{6.38}$$

Let $\theta^-(t)$ denote the negative lobes of $\theta(t)$.

Based on the above preliminaries, we now state and prove an instability criterion for the linear system (2.21).

Theorem 6.51. The system (2.21) has exponentially unstable solutions if (i) $Z(s)$, the 'instability multiplier function', is a function of the complex variable s such that $Z(s)$ is strictly positive real; (ii) $Z(s)G(s+\beta)$ is positive real for some $\beta > 0$ and not positive real for all $\beta \leq 0$; $m(s)q(s+\beta)$ (or $n(s)p(s+\beta)$) and $(\mathrm{Ev}\; q(s+\beta)m(s)p(-s+\beta)n(-s))^{(-)}$ have no common factors on the imaginary axis ; and (iii) with β_i = minimum β of hypothesis (ii),

$$(1/T)\int_{t_o}^{t_o+T} \theta^{-}(t)dt \geq -2\beta_i + \varepsilon \tag{6.39}$$

for some $\varepsilon > 0$, for all $T > 0$.

Proof. As a Lyapunov-Corduneanu function candidate for (6.32) (and hence for (2.21)), choose $V(\underline{x},t)$ defined by (6.35). This quadratic function meets the requirements of Lemma 6.32, and its time-derivative along the solutions of (6.32) satisfies the inequality (6.37). Invoking Corollary 6.322 of Lemma 6.32, we conclude that

$$(1/T)\int_{t_o}^{t_o+T}\left\{\inf_{t \geq t_o} (2\beta,\theta(t)+2\alpha) - \max(\eta_1,\eta_2)\right\} dt \geq \varepsilon \tag{6.40}$$

for some $\varepsilon > 0$, for all $T > 0$ guarantees the exponential instability of the solutions of (6.32). But the linear time invariant system (2.15) has, on the basis of hypothesis (ii) (and the Nyquist stability criterion), solutions with the property

$$\|\underline{x}(t)\|^2 \geq \gamma_o \|\underline{x}(t_o)\|^2 \exp((2\beta_i-\varepsilon_1)(t-t_o)) \tag{6.41}$$

for some positive constants γ_o and ε_1, for all $t \geq t_o$. Hence the use of the integral inequality (6.40) with $\theta(t) \equiv 0$ in the linear time invariant system (2.15) leads to the result that, by an optimal choice of β and α in the generation of quadratic forms by (6.35) and (6.34), we should be able to establish the following relation between the set ($\beta,\alpha,\eta_1,\eta_2$) and β_i :

$$\min(2\beta,2\alpha) - \max(\eta_1,\eta_2) = 2\beta_i \tag{6.42}$$

Note that $\alpha \geq \beta$ and hence, at the cost of apparent generality, we can set $\alpha = \beta$. This facilitates the transformation of the integral inequality (6.40) to (6.39).

The theorem is proved.

Analogous to the geometric interpretation of Theorem 3.42 of Chapter 3, we have the following instability result in which we introduce an improvement over (6.39). This improvement is derivable much as in Sec.3.4 (pp 68-69).

CRITERION 6.51. Let $\beta_i > 0$ be the minimal shift for which the plot of $G(j\omega+\beta_i)$ avoids the negative real axis with ω in $(-\infty,\infty)$. Then the linear time varying system (2.21) is exponentially unstable if $\theta(t)$, the normalized rate of variation of $k(t)$, satisfies the global constraint

$$(1/T)\int_{t_o}^{t_o+T} \theta^-(t)dt \geq -M > -\infty$$

for some constant $M > 0$, for all $t_o \geq 0$ and finite $T > 0$; and

$$\lim_{T\to\infty} (1/T)\int_{t_o}^{t_o+T} \theta^-(t)dt \geq -2\beta_i + \varepsilon$$

for some constant $\varepsilon > 0$.

Example 6.51. The linear time invariant system (2.15) with

$$G(s) = (1/(s^3+2s^2+0.500001s+1)) \qquad (!)$$

is unstable for all constant feedback gains K in $[\varepsilon,\infty)$ with $\varepsilon > 0$. It can be verified that $G(j\omega+\beta_i)$ avoids the negative real axis to the left of the point $(-1/K_\ell)$ for finite K_ℓ for finite β_i with ω in $(-\infty,\infty)$. For arbitrarily large K_ℓ, β_i should be correspondingly large. Therefore, invoking the finite gain version (Exercise 6.10) of Criterion 6.51, we conclude that the linear time varying system (2.21) with G(s) given by (!) is exponentially unstable if (6.39) or the equivalent set in Criterion 6.51 is satisfied with $\theta(t)$ replaced by $\theta_{eq}(t)$ defined as follows :

$$k_{eq}(t) = k(t)/(1-(k(t)/K_\ell)) \; ; \; \theta_{eq}(t) = (dk_{eq}/dt)/k_{eq}(t).$$

Note that for K_ℓ arbitrarily large, β_i is correspondingly large, which implies that no restriction need be imposed on $\theta(t)$. β_i is indicative of the amount of 'negative damping' in the system : The larger the negative damping, the greater is the tendency of the system to become unstable for arbitrary variations of the time varying feedback gain.

6.52. Nonlinear System Instability Analysis.

Analogous to the stability analysis (Sec.3.5) of the nonlinear system, we employ, for its instability analysis, $V(\underline{\chi},t)$ generated from the positive realness of $Z(s)G(s+\beta)$ by the path integral (6.33) along with a suitable integral containing the nonlinear function $\varphi(\cdot)$.

The time-derivative of $V'(\underline{\chi},t)$ defined by (6.33) along the solutions of (6.31) is given by

$$(dV'(\underline{\chi},t)/dt)\Big|_{(6.31)} = 2\beta V'(\underline{\chi}(t),t)-k(t)\varphi(q(D)n(D-\beta)y(t))(q(D)m(D-\beta)y(t)) - (r'(D-\beta)y(t))^2 \qquad (6.43)$$

The term on the right hand side of (6.43) involving the nonlinear function $\varphi(\cdot)$ suggests that we should analyze the conditions for the nonnegativity of the integral

$$I_n = \int_0^t \exp(-\alpha\tau)\varphi(n(D-\beta)\sigma(\tau))m(D-\beta)\sigma(\tau)d\tau \qquad (6.44)$$

for some constant $\alpha \geqslant 0$ and $\varphi(\cdot) \in C$, C_m and C_{mo}. In common with the results of Chapter 3, the choice of the multiplier function, $Z(s) = m(s)/n(s)$, is governed by the class of functions to which $\varphi(\cdot)$ belongs.

Let

$$\Phi(\sigma) = \int_0^\sigma \varphi(w)dw \ ; \quad \delta(\sigma) = (\Phi(\sigma)/\varphi(\sigma)\sigma) \text{ for } \sigma \neq 0 \qquad (6.45)$$

$$\delta_s = \sup_\sigma \delta(\sigma) \ ; \text{ and } \delta_m = \inf_\sigma \delta(\sigma) \ . \qquad (6.46)$$

(Note that δ_m is written in place of δ_i of Chapter 3.5. The reason will become obvious in Lemma 6.53 later below). See Sec.3.5(a) (pp 73-74) for some details on these.

We now present the first lemma on the nonnegativeness of I_n defined by (6.44) with $\varphi(\cdot) \in C$ and a specific $Z(s)$. This sets the pattern for other lemmas involving $\varphi(\cdot) \in C_m$ and C_{mo}.

Lemma 6.52.. Let $m_o(s) = a_o+b_os$ and $n_o(s) = 1$ where $a_o > 0$ and $b_o \geqslant 0$. Then the integral

$$I_o = \int_0^t \exp(-\alpha_o\tau)\varphi(n_o(D-\beta)\sigma(\tau))(m_o(D-\beta)\sigma(\tau))d\tau \quad + b_o\,\Phi(\sigma(0)) \tag{6.47}$$

with $\varphi(\cdot)\in C$ is nonnegative for some $\alpha_o > 0$, for all $t \geq 0$ and $\sigma(\cdot)$ (for which the integral is defined) if

$$\alpha_o\,\delta_m\,b_o + (a_o - b_o\beta) \geq 0. \tag{6.48}$$

Proof. See Appendix 6.4.

Lemma 6.53. Let $m_1(s) = (s+\nu\mu)$ and $n_1(s) = (s+\mu)$ where $\mu > 0$ and $0 \leq \nu < 1$. Then the integral

$$I_1 = \int_0^t \exp(-\alpha_1\tau)\varphi(n_1(D-\beta)\sigma(\tau))(m_1(D-\beta)\sigma(\tau))d\tau \tag{6.49}$$

with $\varphi(\cdot)\in C_m$ is nonnegative for some $\alpha > 0$, for all $t \geq 0$ and $\sigma(\cdot)$ (for which the integral is defined) if

$$(\alpha_1\,\delta_m + \mu\nu - \beta) \geq 0. \tag{6.50}$$

Proof. See Appendix 6.5.

Corollary 6.531. Let $Z_2(s) \triangleq (m_2(s)/n_2(s) = (a_o + b_o s) + \sum_i \gamma_i(s+\nu_i\mu_i)/(s+\mu_i)$ where $a_o > 0$, $\mu_i > 0$; $b_o,\gamma_i \geq 0$ and $0 \leq \nu_i < 1$ for all i (finite). Then the integral

$$I_2 = \int_0^t \exp(-\alpha_2\tau)\varphi(n_2(D-\beta)\sigma(\tau))(m_2(D-\beta)\sigma(\tau))d\tau + b_o\,\Phi(\sigma_1(0)) \tag{6.51}$$

with $\varphi(\cdot)\in C_m$ and $\sigma_1(t) \triangleq n_2(D-\beta)\sigma(t)$, is nonnegative for all $t \geq 0$ and $\sigma(\cdot)$ (for which the integral is defined) for some $\alpha_2 > 0$, if

$$\begin{aligned} &\alpha_2\,\delta_m\,b_o + (a_o - b_o\beta) \geq 0 \\ &\alpha_2\,\delta_m + (\mu_i\nu_i - \beta) \geq 0 \quad \text{for all } i \end{aligned} \tag{6.52}$$

In view of Remark 3.52 (p 75), which is applicable here mutatis mutandis, we are interested in the use of a general positive real multiplier function while establishing conditions for the nonnegativity of I_n defined by (6.44).

Lemma 6.54. Let $Z_3(s) \triangleq (m_3(s)/n_3(s)) = 1 + Z_g(s)$ with $z_g(t)$, the inverse transform of $Z_g(s)$, identically zero for $t < 0$. Then the integral

$$I = \int_0^t \exp(-\alpha\tau)\varphi(n_3(D-\beta)\sigma(\tau))(m_3(D-\beta)\sigma(\tau))d\tau \tag{6.53}$$

with $\varphi(\cdot)\in C_m$ is nonnegative for some $\alpha > 0$, for all $t \geq 0$ and $\sigma(\cdot)$ (for which the integral is defined) if

(i) $z_g(t) \leq 0$ for $t \geq 0$;

$$\text{(ii)} \int_0^\infty |z_g(t)| \exp(\beta-(\alpha/2))t\, dt \leq (1/(2+\delta_s-\delta_m)) \tag{6.54}$$

Proof. See Appendix 6.6.

For the derivation of instability criteria for the system (6.31) (and, hence, for the system (2.1)), we need to define the integral

$$V_n''(\underline{\chi},t) = \exp(\alpha t) \int_0^t \exp(-\alpha\tau)\varphi(n(D-\beta)q(D)\chi(\tau))(m(D-\beta)q(D)\chi(\tau))d\tau \tag{6.55}$$

for some constant $\alpha > 0$.

Let

$$V_n(\underline{\chi},t) = V'(\underline{\chi},t) + k(t)V_n''(\underline{\chi},t) \tag{6.56}$$

where $V'(\underline{\chi},t)$ is defined by (6.33) and $V_n''(\underline{\chi},t)$ by (6.55). The time-derivative of $V_n(\underline{\chi},t)$ defined by (6.56) along the solutions of (6.31) is given by

$$(dV_n(\underline{\chi},t)/dt)\Big|_{(6.31)} = 2\beta V'(\underline{\chi}(t),t)-(r'(D-\beta)y(t))^2+(\theta(t)+\alpha)k(t)V_n''(\underline{\chi}(t),t) \tag{6.57}$$

which satisfies the inequality

$$(dV_n(\underline{\chi},t)/dt)\Big|_{(6.31)} \geq \inf_{t\geq 0} (2\beta-\eta_1,\ \theta(t)+\alpha)V_n(\underline{\chi}(t),t) \tag{6.58}$$

where η_1 is given by (6.38).

Note that $V_n(\underline{\chi},t)$, defined by (6.56) with $V_n(\underline{\chi},t)$ constructed to be nonnegative, satisfies inequality (6.21) of Corollary 6.321 with $\underline{x}$ replaced by $\underline{\chi}$.

We now prove an instability criterion for the system (6.31) (and, hence, for the system (2.1))with $\varphi(\cdot)\in C$.

Theorem 6.52. The system (2.1) with $\varphi(\cdot)\in C$ and under Assumption U1 (Sec.2.32, pp 30-31) has solutions with property (6.22) for some $\alpha > 0$ if there exists a multiplier function $Z(s) = a_o + b_o s$ with $a_o > 0$ and $b_o \geq 0$ such that

(i) $\operatorname{Re}(a_o+ b_o j\omega)G(j\omega+\beta) \geq 0$ for some $\beta > 0$ (obtained from Assumption U1 and (6.30)), for all real ω ; there are no common factors between $(a_o+ b_o s)q(s+\beta)$ (or $p(s+\beta)$) and $(Ev(a_o+ b_o s)q(s+\beta)p(-s+\beta))^{(-)}$ on the imaginary axis ;

(ii) there exists an $\alpha_o > 0$ satisfying the inequality

$$(\alpha_o \delta_m - \beta) b_o + a_o \geq 0 \quad ; \tag{6.48}$$

(iii) the integral inequality

$$(1/T) \int_{t_o}^{t_o+T} (\theta(t) + \alpha_o - 2\beta + \eta_1)^{-} dt \geq \alpha_o - 2\beta + \eta_1 - \varepsilon \tag{6.59}$$

holds for some α_o of hypothesis (ii), $\varepsilon > 0$ and for all $T > 0$, where η_1 is given by (6.38) with $m(s) = (a_o + b_o s)$ and $n(s) = 1$.

Proof. As a Lyapunov-Corduneanu function candidate for the system (6.31), choose

$$V_{n1}(\underline{x}, t) = V'(\underline{x}, t) + k(t) \; (V_n''(\underline{x}, t) + \exp(\alpha t) b_o \Phi(\sigma(0)) \tag{6.60}$$

where $V'(\underline{x},t)$ and $V_n''(\underline{x},t)$ are as defined by (6.33) and (6.55) respectively with $m(s) = (a_o + b_o s)$ and $n(s) = 1$. Note that $\underline{\chi}$ in (6.33) and (6.55) has been replaced by $\underline{x}$ because $\tilde{n}(s) = 1$. By virtue of hypotheses (i) and (ii), $V_{n1}(\underline{x},t)$ has the property (6.21) of Corollary 6.322 with α replaced by α_o obtained from (6.48). See Lemma 6.52.

The time-derivative of $V_{n1}(\underline{x},t)$ along the solutions of (6.31) (and hence of (2.1)) satisfies the inequality (6.58) where η_1 is given by (6.38) with $m(s) = a_o + b_o s$ and $n(s) = 1$, and α is replaced by α_o. Invoke Corollary 6.322 to conclude that the global integral inequality (6.59) guarantees property (6.22) for the solutions of (2.1). The theorem is proved.

Remark 6.51. Suppose Theorem 6.52 is applied to the linear time invariant system (2.15), we can at best conclude that the solutions of (2.15) have the property

$$\| \underline{x}(t) \| \geq \eta \; \| \underline{x}(t_o) \| \; \exp\,(2\beta_i - \varepsilon)(t - t_o)$$

for some $\varepsilon > 0$ and $\eta > 0$, for all $t \geq t_o \geq 0$. Hence an optimum choice of $\alpha_o = 2\beta_i$ and, at best, $2\beta - \eta_1 = 2\beta_i - \varepsilon_1$ for some $\varepsilon_1 > 0$.

As regards the geometric interpretation of Theorem 6.52, the construction of Sec.3.5 (p 79) is applicable with the modification that $G(j\omega + \beta)$ is to be used in the place of $G(j\omega)$. There is no loss of generality in assuming that a_o is equal to 1. The slope of the Popov line along with the inequality (6.48) and the chosen value of β gives the lower bound on α_o. η_1 is evaluated from (6.38) with $m(s) = (1 + b_o s)$ and $n(s) = 1$. Satisfaction of the bound (6.59), which could be improved to the form of (3.62) (p 79) as in Sec.3.4, then guarantees property (6.22) for the solutions of (2.1) under Assumption U1 (Sec.2.32, pp 30-31).

Theorem 6.53. The system (2.1) with $\varphi(\cdot) \in C_m$ and under Assumption U1 (Sec. 2.32) has solutions with property (6.22) for some $\alpha > 0$ if there exists a multiplier function

$$Z_2(s) \triangleq (m_2(s)/n_2(s)) = a_o + b_o s + \sum_i \gamma_i (s + \nu_i \mu_i)/(s + \mu_i), \quad (6.61)$$

where constants $a_o > 0$, $\mu_i > 0$; b_o, $\gamma_i \geq 0$, $0 \leq \nu_i < 1$ for all i finite, such that

(i) $\text{Re}\ Z_2(j\omega)G(j\omega + \beta) \geq 0$

for some $\beta > 0$ (obtained from Assumption U1 and (6.30)) for all real ω ; $m_2(s)q(s+\beta)$ (or $n_2(s)p(s+\beta)$) and $(\text{Ev}\ m_2(s)q(s+\beta)p(-s+\beta)n_2(-s))^{(-)}$ have no common factors on the imaginary axis ;

(ii) there exists an $\alpha_1 > 0$ satisfying the set of inequalities

$$(\alpha_1 \delta_m + \mu_i \nu_i - \beta) \geq 0 \qquad \text{for all finite } i,$$

$$(\alpha_1 \delta_m - \beta) b_o + a_o \geq 0 \qquad ;$$

(iii) the integral inequality (6.59) holds with α_o replaced by α_1 of hypothesis (ii), and η_1 is given by (6.38) with $(m_2(s)/n_2(s)$ defined in (6.61) .

Proof. Similar to the proof of Theorem 6.52 but using Corollary 6.531.

Theorem 6.54. The system (2.1) with $\varphi(\cdot) \in C_m$ and under Assumption U1 (Sec. 2.32) has solutions with property (6.22) for some $\alpha > 0$, if there exists a multiplier function

$$Z_3(s) = (m_3(s)/n_3(s)) = 1 + Z_g(s) \qquad (6.62)$$

with $z_g(t)$, the inverse transform of $Z_g(s)$, identically zero for $t \leq 0$, such that

(i) $\text{Re}\ Z_3(j\omega)G(j\omega + \beta) \geq 0$

for some $\beta > 0$ (obtained from Assumption U1 and (6.30)), for all real ω ; $m_3(s)q(s+\beta)$ (or $n_3(s)p(s+\beta)$) and $(\text{Ev}\ m_3(s)q(s+\beta)p(-s+\beta)n_3(-s))^{(-)}$ have no factors on the imaginary axis ;

(ii) $z_g(t) \leq 0$ for $t \geq 0$ and, for some $\alpha_2 > 0$,

$$\int_0^{\infty} |z_g(t)| \exp(\beta-(\alpha_2/2))t \, dt \leq 1/(2 + \delta_s - \delta_m) \quad ;$$

(iii) the integral inequality (6.59) holds with α_o replaced by α_2 of hypothesis (ii), and η_1 is given by (6.38) with $m_3(s)/n_3(s)$ defined in (6.62).

We now compare the instability criteria obtained in Sec.6.4 with those of the present section.

Table 6.3

Comparison of Instability Criteria. Assumption U1 (Sec.2.32,pp 30-31)

No.	System under consideration	Lemma used for deriving the instability criteria. Multiplier function.	Main instability conditions on G(s).	Instability constraint on k(t).	Remarks.
1	(2.21) Linear time varying	Lyapunov-Chetaev (Lemma 6.31). $Z(s)=1$.	Re $G(j\omega) \geq 0$ for all real ω. G(s) is not positive real.	No constraint..	Real part requirement on G(s) is too restrictive. Nyquist's instability criterion not obtained as a special case.
2	,,	Lyapunov-Corduneanu (Lemma 6.3 6.32) Z(s) =arbitrary strictly positive real function.	(i) $Z(s)G(s+\beta)$ positive real for some $\beta > 0$, not positive real for any $\beta \leq 0$; (ii) Z(s) strictly positive real.	Global lower bound on $\theta^-(t)$, dependent on the negative damping in the system.	Least restrictive set of instability conditions. Nyquist's instability criterion is obtained as a special case.
3	(2.1) Nonlinear time varying. $\varphi \in C$	Lyapunov-Chetaev (Lemma 6.31). $Z(s) = a_o+b_os$.	(i) Re $(a_o+b_oj\omega)$ $G(j\omega) \geq 0$ for all real ω; (ii) G(s) (a_o+b_os) is not positive real.	Global upper bound on $\theta^+(t)$ depending on δ_s.	Geometric interpretation similar to Criterion 3.51 (p 79).

No.	System under consider-ation.	Lemma used for deriving the instability criteria. Multiplier function.	Main instability conditions on G(s)	Instability constraint on k(t).	Remarks
4	(2.1)Non-linear time varying. $\varphi(\cdot)\in C$.	Lyapunov-Corduneanu (lemma 6.32). $Z(s) = a_o+b_os$	$Re(a_o+b_oj\omega)$ $G(j\omega+\beta)\geq 0$ for some $\beta>0$, for all real ω.	Global lower bound on $\theta^-(t)$ depending on δ_m.	Geometric interpretation similar to Criterion 3.51 (p 79). Counterpart of Criterion 3.51.
5	(2.1)Non-linear time varying. $\varphi(\cdot)\in C_m$	Lyapunov-Chetaev (Lemma 6.31). Not known how to choose Z(s)	—	—	—
6	,,	Lyapunov-Corduneanu (Lemma 6.32). Multiplier function as in Chapter 3.	Re $Z(j\omega)G(j\omega+\beta)\geq 0$ for some $\beta>0$, for all real ω. Certain restriction on Z(s).	Global lower bound on $\theta^-(t)$.	Geometric interpretation similar to those of Chapter 3 (Sec.3.5), p 79, with $G(j\omega+\beta)$ replacing $G(j\omega-\beta)$.

Remark 6.52. Suppose we attempt derivation of instability criteria for the systems (2.1) and (2.21) under Assumption S1 (Sec.2.32, pp 30-31), we find that β of Basic Result 6.51 is now negative. Consequently, the linear system (2.21) and the nonlinear system (2.1) have solutions, on the basis of Theorems 6.51 and 6.52 (-6.54), with the properties

$$\|\underline{x}(t)\|^2 \geq \gamma_0\|\underline{x}_0\|^2 \exp(-\gamma_1(t-t_o)), \qquad t\geq t_o\geq 0$$

and

$$\int_{t_o}^{t} \exp(\gamma_2\tau)\|\underline{x}(\tau)\|^2 \, d\tau \geq \gamma_3\|\underline{x}_0\|^2 (\exp(-\gamma_4(t-t_o))-\exp(-\gamma_5(t-t_o)))$$

respectively, for some positive constants γ_i, $i = 0,1,\cdots,5$. Details regarding the proof of this conclusion are left as an exercise.

As explained in Sec.6.42 (Case 1), the Lyapunov- Chetaev lemma is inapplicable under Assumption S1 when used in the way we are accustomed to. See below for a reference to the results of other authors.

6.6. Comparison with other results in the literature.

(a) Skoog [S 5] , Steding and Bergen [S 10] , Takeda and Bergen [T 1(a)] , and others presuppose an unstable G(s). Skoog further requires that Re $G(j\omega)$ ≥ 0 for all real ω.

(b) Takeda and Bergen [T 1(a)] state an instability condition of the form $(1 + j\omega\mu)G(j\omega) \geq \delta > 0$, for some constant $\mu > 0$ for all real ω , for the nonlinear time invariant system (2.24). Suppose G(s) does not have poles in the right half of the complex plane (Example 6.51), the criteria of Takeda and Bergen, Steding and Bergen do not apply. Further, even when they are applicable, the instability sector for the feedback gain may be significantly smaller than the Hurwitz instability sector (Exercise 6.12).

(c) The instability criteria of Noldus [N 3(b)(c)] , obtained from an application of the Lyapunov-Chetaev instability lemma, deal with two forms of instability : (i) A type of instability which drives the system in a sustained bounded oscillation, called weakly unstable by Noldus. In this case, the Lyapunov-Chetaev lemma is to be used with the knowledge of the boundary $\|\underline{x}\| < h$. After establishing instability as in Sec.6.4, it is to be shown that the solutions never leave the boundary again. (ii) A type of instability pertaining to the existence of unbounded oscillations, called strongly unstable by Noldus. This corresponds to an application of the Lyapunov-Corduneanu lemma here in Sec.6.5. When ε of the solution property (6.22) is set to zero, an application of Corollary 6.322 also yields some information on the possibility of bounded self-excited oscillations. Further details are to be worked out.

The criterion of Noldus as applied to the nonlinear time invariant system (2.24), under the usual assumptions (Sec,2.2, p 23),reads as follows : The system (2.24) with $\varphi(\cdot)$ is strongly unstable if

(6.63)

(i) $1 + (d\varphi(\sigma)/d\sigma) \Big|_{\sigma=0} G(s-r)$ (6.63)

has $n_1 \geqslant 1$ zeros in the open right half plane and all other zeros in $\mathrm{Re}\, s \leqslant -\varepsilon < 0$, for a scalar $r < 0$; and

(ii) $\mathrm{Re}(1-\alpha(j\omega - r))(G(j\omega - r))^{\pm 1} - \alpha r \geqslant \eta^2 > 0$ (6.64)

for a scalar $\alpha \leqslant 0$, for all real ω.

Theorem 6.52 on specialization to (2.24) guarantees solutions with property (6.22) for some $\alpha > 0$ if

(i) $\mathrm{Re}(a_o + b_o j\omega)G(j\omega + \beta) \geqslant 0$

for some $\beta > 0$ (obtained from Assumption U1 and (6.30)), for all real ω ; there are no common factors between $(a_o + b_o s)q(s+\beta)$ (or $p(s+\beta)$) and $(\mathrm{Ev}(a_o + b_o s)q(s+\beta)p(-s+\beta))^{(-)}$ on the imaginary axis ;

(ii) β in hypothesis (i) can be so chosen that $(2\beta - \eta_1) > 0$ where η_1 is given by (6.38) with $m(s) = (a_o + b_o s)$ and $n(s) = 1$.

Remark 6.61. Note that Theorem 6.52 makes no reference to the slope of $\varphi(\sigma)$ at $\sigma = 0$. See Exercise 6.13.

As regards the instability of the nonlinear time varying system (2.1), Noldus gives a constraint on the rate of variation of $k(t)$ which is local in character. It is not obvious in the contribution of Noldus how a general multiplier function can be employed to derive instability criteria.

(d) The instability circle criterion of Brockett and Lee [B 18] does not reduce to the Nyquist instability criterion when applied to the linear time invariant system (2.15). Further, their Popov-type instability criterion for the time-invariant nonlinear system (2.24) requires an unstable G(s).

The above comment holds for the instability criteria of Davis [D 3(a)] derived for the linear time invariant system (2.21). See Sec.3.4(b) (p 72) for comments on the stability criteria of Davis which apply, mutatis mutandis, to the instability criteria.

6.7. Conclusions.

Some instability criteria are derived for linear and nonlinear time varying systems described by (2.21) and (2.1) respectively. These criteria, which are based on the instability lemmas of Lyapunov-Chetaev and Lyapunov-Corduneanu, are better treated as tentative, needing further clarification and substantial improvement.

The instability results obtained from an application of the Lyapunov-Corduneanu lemma have the desirable feature that the Nyquist criterion for the linear time- invariant system (2.15) is obtained as a special case.

The contents of the chapter are believed to be inadequate since some basic instability problems (Table 6.2) are still unsolved. Further, Lemma 6.54, which is supposed to be the counterpart of Lemma 3.53 and is needed in the proof of Theorem 6.54, contains $(1/(2+\delta_s-\delta_m))$ as the upper bound on the time-domain integral of the instability multiplier function. This is in contrast with $(1/(1+\delta_s-\delta_m))$ appearing in Lemma 3.53 for the time domain integral of the stability multiplier function. The question is how to improve upon the multiplier function time-domain constraint in Lemma 6.54.

The instability criteria of the chapter are applicable to the case of a periodic k(t). But, contrary to intuition, the instability conditions for the systems (2.1) and (2.21) obtained from these criteria do not contain the period of T explicitly. Hence the criteria need modification to deal with periodic k(t) .

Another problem of interest is the prediction of the existence, amplitude and frequency of oscillations of the system (2.1) in the case of local instability of its solutions.

In the next chapter, an application of the converse Schwarz inequality to the L_2-instability analysis of feedback systems holds promise of more general instability criteria.

Appendix 6.1

Proof of Lemma 6.32. Integration of the differential inequality (6.18) gives

$$v_2(\underline{x}(t),t) \geqslant v(\underline{x}_o,t_o)\exp\int_{t_o}^{t}\lambda(\tau)d\tau \tag{6.65}$$

Consequently, from (6.17) and (6.65), we get

$$\gamma_1 \|\underline{x}(t)\|^2 \geq \gamma_o \|\underline{x}_o\|^2 \exp\left(\int_{t_o}^{t} \lambda(\tau)d\tau\right) \tag{6.66}$$

from which follows the inequality (6.19) with $\eta_o = (\gamma_o/\gamma_1)^{(1/2)}$. The lemma is proved.

Appendix 6.2

Proof of Corollary 6.321. In the place of (6.66), we now have

$$\gamma_1 \|\underline{x}(t)\|^2 + \gamma_2 \exp(\alpha t) \int_{t_o}^{t} \exp(-\alpha\tau) \|\underline{x}(\tau)\|^2 d\tau$$

$$\geq \gamma_o \|\underline{x}_o\|^2 \exp\left(\int_{t_o}^{t} \lambda(\tau)d\tau\right), \quad t \geq t_o \geq 0 \tag{6.67}$$

If $(1/T)\int_{t_o}^{t_o+T} \lambda(\tau)d\tau \geq \alpha-\varepsilon$ for some $\varepsilon > 0$, for all $T > 0$, then from (6.67),

$$\gamma_1 \|\underline{x}(t)\|^2 \exp(-\alpha t) + \gamma_2 \int_{t_o}^{t} \exp(-\alpha\tau) \|\underline{x}(\tau)\|^2 d\tau$$

$$\geq \gamma_o \|\underline{x}_o\|^2 \exp(-\varepsilon(t-t_o)), \quad t \geq t_o \geq 0. \tag{6.68}$$

Let

$$U(t) \triangleq \int_{t_o}^{t} \exp(-\alpha\tau) \|\underline{x}(\tau)\|^2 d\tau .$$

Then (6.68) can be written as

$$\gamma_1 (dU/dt) + \gamma_2 U(t) \geq \gamma_o \|\underline{x}_o\|^2 \exp(-\varepsilon(t-t_o)), \quad t \geq t_o \geq 0 \tag{6.69}$$

or, by virtue of the fact that $U(t_o) = 0$,

$$U(t) \geq (\gamma_o/\gamma_1) \|x_o\|^2 \int_{t_o}^{t} \exp(-\gamma_2(t-\tau)/\gamma_1) \exp(-\varepsilon(\tau-t_o))\, d\tau$$

$$= (\gamma_o/(\gamma_2-\gamma_1\varepsilon)) \|\underline{x}_o\|^2 (\exp(-\varepsilon(t-t_o)) - \exp(-\gamma_2(t-t_o)/\gamma_1))$$

from which we conclude that

$$\int_{t_o}^{t} \exp(-\alpha\tau) \|\underline{x}(\tau)\|^2 d\tau \geq \gamma_3 \|\underline{x}_o\|^2 (\exp(-\varepsilon(t-t_o)) - \exp(-\gamma_2(t-t_o)/\gamma_1)) \tag{6.22}$$

for all $t \geq t_o \geq 0$, where $\gamma_3 = \gamma_o/(\gamma_2 - \gamma_1\varepsilon)$.

Appendix 6.3

Proof of Lemma 6.51. If $\phi(t)$ is a solution of (6.31), then $n(D-\beta)\phi(t)$ is a solution of (2.1). If there is a solution of (6.31) which increases exponentially, so do $\phi(t)$ and its first $(\eta + n - 1)$ derivatives, Denoting $n(D-\beta)\phi(t)$ by $\phi'(t)$, we conclude that $\phi'(t)$ and its first $(n - 1)$ derivatives increase exponentially, provided that $n(0) \neq 0$. This proves the lemma.

Appendix 6.4

Proof of Lemma 6.52. We have

$$I_o = (a_o - b_o\beta)\int_0^t \exp(-\alpha_o\tau)\ \varphi(\sigma(\tau))(\sigma(\tau)\ d\tau + b_o\int_0^t \exp(-\alpha_o\tau)\varphi(\sigma(\tau))(d\sigma/d\tau)d\tau + b_o\Phi(\sigma(0)) \quad (6.70)$$

Integrate by parts the second integral on the right hand side of (6.70) to get

$$I_o = (a_o - b_o\beta)\int_0^t \exp(-\alpha_o\tau)\varphi(\sigma(\tau))\sigma(\tau)\ d\tau + b_o \exp(-\alpha_o t)\Phi(\sigma(t)) + \alpha_o b_o\int_0^t \exp(-\alpha_o\tau)\Phi(\sigma(\tau))\ d\tau \ .$$

But $\Phi(\sigma) \geq \delta_m\varphi(\sigma)\sigma$ for all $\sigma \neq 0$. Hence I_o is nonnegative if, for some $\alpha_o > 0$,

$$(\alpha_o b_o \delta_m + a_o - b_o\beta) \geq 0\ . \quad (6.48)$$

The lemma is proved.

Appendix 6.5

Proof of Lemma 6.53. We have

$$(m_1(s-\beta)/n_1(s-\beta)) = (s-\beta+\nu\mu)/(s-\beta+\mu) = 1 - (\mu(1-\nu)/(s-\beta+\mu)).$$

Let
$$\sigma_1(t) \triangleq (D+\mu-\beta)\sigma(t) \quad (6.71)$$
and
$$\sigma_2(t) = \mu(1-\nu)\int_0^t \exp\left((\beta-\mu)(t-\tau)\right)\ \sigma_1(\tau)\ d\tau\ . \quad (6.72)$$

Then $(d\sigma_2/dt) = \mu(1-\nu)\sigma_1(t) + (\beta-\mu)\sigma_2(t)$

or
$$\sigma_1(t) = (1/\mu(1-\nu))((\mu-\beta)\sigma_2(t) + (d\sigma_2/dt)) \quad (6.73)$$

Hence, from (6.49), we get

$$I_1 = \int_0^t \exp(-\alpha_1\tau)\ \varphi(\sigma_1(\tau))(\sigma_1(\tau) - \sigma_2(\tau))d\tau = \int_0^t \exp(-\alpha_1\tau)(\varphi(\sigma_1(\tau)) - \varphi(\sigma_2(\tau)))(\sigma_1(\tau)-\sigma_2(\tau))d\tau + \int_0^t \exp(-\alpha_1\tau)\varphi(\sigma_2(\tau))(\sigma_1(\tau)-\sigma_2(\tau))d\tau \quad (6.74)$$

The first integral on the right hand side of (6.74) is nonnegative by virtue of the fact that $\varphi(\cdot)\in C_m$. The second integral on the right hand side of (6.74) can be written , using (6.73), as

$$\int_0^t \exp(-\alpha_1\tau)\varphi(\sigma_2(\tau))(\sigma_1(\tau)-\sigma_2(\tau))d\tau = (1/\mu(1-\nu))\int_0^t \exp(-\alpha_1\tau)\varphi(\sigma_2(\tau))((-\beta+\mu\nu)\sigma_2(\tau) + (d\sigma_2/d\tau))d\tau \qquad (6.75)$$

which is of the form considered in Lemma 6.52. Hence the second integral on the right hand side of (6.74) is nonnegative if $0\leq\nu<1$ and

$$\alpha_1\delta_m + \mu\nu - \beta \geq 0 \qquad (6.50)$$

Appendix 6.6

Proof of Lemma 6.54. Let $n_3(D-\beta)\sigma(\tau) \triangleq \sigma_1(\tau)$. We have

$$I = \int_0^t \exp(-\alpha\tau)\varphi(\sigma_1(\tau))\,\sigma_1(\tau) + \int_0^t \exp(-\alpha\tau)\varphi(\sigma_1(\tau))(\int_0^\infty z_g(\tau')\exp(\beta\tau')\sigma_1(\tau-\tau')d\tau')d\tau \qquad (6.76)$$

Assuming that the order of integration in the last term of (6.76) can be interchanged, we get (after some obvious manipulations)

$$I = \int_0^t \exp(-\alpha\tau)\varphi(\sigma_1(\tau))\sigma_1(\tau)d\tau + \int_0^\infty z_g(\tau')\exp((\beta-(\alpha/2))\tau')(\int_0^t \exp(-\alpha\tau/2)\varphi(\sigma_1(\tau))\sigma_1(\tau-\tau')\exp(-\alpha(\tau-\tau')/2)d\tau)d\tau' \qquad (6.77)$$

From the monotone property of $\varphi(\cdot)$ we have

$$\varphi(y_1)(y_1-y_2) \geq \Phi(y_1) - \Phi(y_2) \qquad \text{for all } y_1 \text{ and } y_2 \qquad (6.78)$$

Define $y_1 = \sigma_1(\tau)$ and $y_2 = \sigma_1(\tau-\tau')\exp(-\alpha(\tau-\tau')/2)$. Using (6.78), we can write the following inequality :

$$\int_0^t \exp(-\alpha\tau/2)\varphi(\sigma_1(\tau))(\sigma_1(\tau)-\sigma_1(\tau-\tau')\exp(-\alpha(\tau-\tau')/2))d\tau \geq \int_0^t \exp(-\alpha\tau/2)\Phi(\sigma_1(\tau))d\tau - \int_0^t \exp(-\tau\alpha/2)\Phi(\sigma_1(\tau-\tau')\exp(-\alpha(\tau-\tau')/2))d\tau \qquad (6.79)$$

The last integral of (6.79) can be written as follows by changing the variable of integration to $\tau_1 = \tau - \tau'$:

$$\int_0^t \exp(-\alpha\tau/2)\Phi(\sigma_1(\tau-\tau')\exp(-\alpha(\tau-\tau')/2))d\tau = \int_{-\tau'}^{t-\tau'} \exp(-\alpha(\tau_1+\tau')/2)\,\Phi(\sigma_1(\tau_1)\exp(-\alpha\tau_1/2))d\tau_1 .$$

But $\Phi(\sigma_1(\tau_1)\exp(-\alpha\tau_1/2)) \leq \delta_s\varphi(\sigma_1(\tau_1)\exp(-\alpha\tau_1/2))\sigma_1(\tau_1)\exp(-\alpha\tau_1/2)$, and

$\Phi(\sigma_1(\tau_1)) \geq \delta_m \varphi(\sigma_1(\tau_1))\sigma_1(\tau_1)$.

Therefore, noting that $\tau' \geq 0$ and that $\sigma_1(\tau_1) = 0$ for $\tau_1 < 0$, the right hand side of the inequality (6.79) is greater than or equal to

$$\delta_m \int_0^t \exp(-\alpha\tau/2)\varphi(\sigma_1(\tau))\sigma_1(\tau)d\tau - \delta_s \int_0^t \exp(-\alpha\tau)\exp(-\alpha\tau'/2)\varphi(\sigma_1(\tau)\exp(-\alpha\tau/2))\sigma_1(\tau)d\tau .$$

Hence, from (6.79) we get

$$\int_0^t \exp(-\alpha\tau)\varphi(\sigma_1(\tau))\sigma_1(\tau-\tau')\exp(\alpha\tau'/2)d\tau \leq \delta_s \int_0^t \exp(-\alpha\tau)\exp(-\alpha\tau'/2)\varphi(\sigma_1(\tau)\exp(-\alpha\tau/2))\sigma_1(\tau)d\tau - \delta_m \int_0^t \exp(-\alpha\tau/2)\varphi(\sigma_1(\tau))\sigma_1(\tau)d\tau + \int_0^t \exp(-\alpha\tau/2)\varphi(\sigma_1(\tau))\sigma_1(\tau)d\tau \qquad (6.80)$$

But the first integral on the right hand side of (6.80) satisfies the inequality,

$$\delta_s \int_0^t \exp(-\alpha\tau)\exp(-\alpha\tau'/2)\varphi(\sigma_1(\tau)\exp(-\alpha\tau/2))\sigma_1(\tau)d\tau \leq \delta_s \int_0^t \exp(-\alpha\tau)\varphi(\sigma_1(\tau))\sigma_1(\tau)d\tau,$$

the second integral, the inequality

$$-\delta_m \int_0^t \exp(-\alpha\tau/2)\varphi(\sigma_1(\tau))\sigma_1(\tau)d\tau \leq -\delta_m \int_0^t \exp(-\alpha\tau)\varphi(\sigma_1(\tau))\sigma_1(\tau)d\tau,$$

and the third integral, the inequality

$$\int_0^t \exp(-\alpha\tau/2)\varphi(\sigma_1(\tau))\sigma_1(\tau)d\tau = \int_0^t \exp(-\alpha\tau)\varphi(\sigma_1(\tau))\sigma_1(\tau)d\tau + \int_0^t (\exp(-\alpha\tau/2)-\exp(-\alpha\tau))\varphi(\sigma_1(\tau))\sigma_1(\tau)d\tau \leq 2\int_0^t \exp(-\alpha\tau)\varphi(\sigma_1(\tau))\sigma_1(\tau)d\tau .$$

Therefore, the right hand side of (6.80) can be replaced by

$$(2 + \delta_s - \delta_m)\int_0^t \exp(-\alpha\tau)\varphi(\sigma_1(\tau))\sigma_1(\tau)d\tau$$

which in combination with (6.77) enables us to conclude that I defined by (6.76) is nonnegative if $z_g(t) \leq 0$ for all $t \geq 0$, and inequality (6.54) is satisfied. The lemma is proved.

EXERCISES

6.1. Consider the system (2.21) with $G(s) = (3s-4)/(s^2-2s+2)$. Write the equations of the adjoint system and indicate whether instability conditions are any more easily derivable.

6.2. Consider the system (2.21) with $G(s) = 1/s(s-2)$. Apply the circle stability criterion to its adjoint system and hence state the conditions for instability of the original system. Also analyse the stability of the system obtained from the original system by reversing the sign of t.

6.3. Show that the linear system (6.5) is unstable if the integral inequality (6.5) is satisfied. What would be your conclusions on the instability of (6.5) if $q_{n-1} = 0$ and $p_{n-1} > 0$?

6.4. Consider the lightly damped Mathieu equation

$$(d^2 x/dt^2) + 0.001(dx/dt) + (a+b \cos 2t)x = 0$$

with the constants a and b so chosen as to satisfy Assumption S1 (Sec.2.32,p 30). For suitable combinations of a and b, the equation has unstable solutions [M 7] . Apply the classical methods of instability analysis given in Table 6.1 and comment on the results so obtained.

6.5. In [C 7] , we find a linear time varying system described by (2.21) with

$$G(s) = K_1(s+0.3)/(s^3+3.5s^2+(K_1-4.5)s+0.3K_1)$$

where $K_1 > 4.92$. Based on analog computer studies [C 7], it is stated that the time invariant system (2.15) with the above G(s) and $K_1 > 4.92$ is unstable for all K in $[0,\infty)$, but the linear time varying system (2.21) with $k(t) = K_2 \cos 4t$ becomes unstable for some suitable choice of K_2 and K_1 , the latter being greater than 4.92. How can one verify this assertion analytically ?

6.6. Apply the instability criterion (6.8) to the systems of Exercises 6.1 and 6.2 and compare with the results obtainable from the integral inequality (6.7).

6.7. Prove that $V_1(\underline{x},t)$ and $V_2(\underline{x},t)$ defined by (6.24) and (6.25) respectively are sign indefinite, provided that $Z(s-\alpha)$ and $Z(s-\beta)G(s-\beta)$ are not positive real but the real part conditions are satisfied.

6.8. In order to generate sign indefinite quadratic forms for the system (2.1) or (2.21) under Assumption U1 (p 31), we find that the requirement Re $Z(j\omega) \geq 0$ is violated. Suppose for the same problem Re $Z(j\omega+\alpha) \geq 0$ for some α for all real ω . Analyze how this could be used to generate a sign indefinite quadratic form. Also note the corresponding changes to be made in using the inequality Re $Z(j\omega)G(j\omega) \geq 0$ for all real ω, for a similar generation of the sign indefinite quadratic form.

6.9. (a) Using the method of Sec.6.4, generate sign indefinite quadratic forms for the system (2.21) with G(s) given in Exercises 6.2 and 6.3.

(b) Employing $V_{c3}(\underline{x},t)$ defined by (6.28), derive instability conditions for the system (2.1) with G(s) of part (a).

6.10. Derive the finite gain version of Theorems 6.41, 6.42 and 6.51 and of the Instability Criterion 6.51.

6.11. For the G(s) of Exercise 6.2, apply the criterion of Takeda and Bergen [T 1(a)] and determine the sector in which the nonlinearity should lie for instability. Compare with the Hurwitz instability sector and also with the results of Theorem 6.52.

6.12. (a) Apply the criterion of Noldus [N 3(b)] to the system (2.1) with $k(t) \equiv$ a constant and $G(s) = 1/(s+1)^3$. Compare with the result obtainable from an application of Theorems 6.52 and 6.42; (b) Also consider the time varying system (2.1) and repeat part (a); (c) Repeat parts (a) and (b) for $G(s) = 1/(s+1)^2 (s+2)$.

6.13. Skoog and Lau [S 6] derive an instability criterion, using the Lyapunov-Chetaev lemma for slowly varying systems described by

$$(d\underline{x}/dt) = A(t)\underline{x} \quad . \tag{6.81}$$

Their criterion reads : If the matrix A(t) has some eigenvalues in the right half plane and all the eigenvalues are bounded away from the imaginary axis, then then if $\sup_{t\geq 0}\|(dA(t)/dt)\|$ is sufficiently small, the system (6.81) has unbounded solutions. The criterion contains the restriction that the eigenvalues' of A(t) should not cross the imaginary axis.

(a) Apply the above criterion to the scalar first order system

$$(dy/dt) = k(t)y \tag{6.82}$$

where k(t) can take values in $(-\infty,\infty)$. The system (6.82) has an explicit solution which can be written in closed form. Comment on the usefulness of the Skoog and Lau criterion.

(b) Repeat part (a) for the system considered in [S 6] :

$$(dy_1/dt) = (-1+k_1(t))y_1 + k_2(t)y_2 \; ; \; (dy_2/dt) = k_2(t)y_1 + (-1+k_1(t))y_2 \, ,$$

whose exact solution, given by

$$y_1(t) = (1/2)\exp(-(t-t_o) + \int_{t_o}^{t} k_1(\tau)d\tau)(\exp(\int_{t_o}^{t} k_2(\tau)d\tau)(y_1(t_o)+y_2(t_o)) + \exp(-\int_0^t k_2(\tau)d\tau)(y_1(t_o)-y_2(t_o))),$$

$$y_2(t) = (1/2)\exp(-(t-t_o) + \int_{t_o}^{t} k_1(\tau)d\tau)(\exp(\int_{t_o}^{t} k_2(\tau)d\tau)(y_1(t_o)+y_2(t_o)) + \exp(-\int_0^t k_2(\tau)d\tau)(y_2(t_o)-y_1(t_o))),$$

enables us to conclude that the system is unstable for all $k_1(t)$ and $k_2(t)$ if, for some $\varepsilon > 0$,

$$(1/T)\int_{t_o}^{t_o+T} k_1(t)dt \geq 1 + \varepsilon + (1/T)\int_{t_o}^{t_o+T} |k_2(t)|\, dt$$

for all $T > 0$. Do the eigenvalue variations about the imaginary axis play any role in the instability criterion ?

(c) The extension given in [S 6] of the instability criterion stated above to the nonlinear system

$$(d\underline{x}/dt) = A(t)\underline{x} + \underline{f}(\underline{x},t) \tag{6.83}$$

is based on the assumption that $(\|\underline{f}(\underline{x},t)\| / \|\underline{x}\|) \rightarrow 0$ as $\|\underline{x}\| \rightarrow 0$. Compare with the possible extensions of the results of the chapter to the system (6.83) without the smallness restriction on $\underline{f}(\underline{x},t)$.

6.14. Apply the instability criterion of Davis [D 3] to the linear time varying system (2.21) with $G(s) = 1/(s^2-0.001s+c)$ for some constant $c > 0$, and $k(t) \geq 0$ for all $t \geq 0$. Compare with those obtainable from Theorems 6.41 and 6.51, and from the classical methods of Table 6.1.

6.15. Suppose $\varphi(\cdot) \in C_{mo}$ and $Z(s) \triangleq (n(s)/m(s)) = (s+\mu\nu)/(s+\mu)$ with $\nu \geq 1$ and $\mu > 0$. Using the odd monotonicity property

$$\varphi(\sigma_1)(\sigma_1 + \sigma_2) - \varphi(\sigma_2)(\sigma_1 - \sigma_2) \geq 0$$

for all σ_1 and σ_2, find conditions for the nonnegativity of the functional (6.44). Generalize the result to the use of the multiplier function

$$Z(s) \triangleq (m(s)/n(s)) = a_o + b_o s + \sum \gamma_i (s + {}_i\mu_i)/(s+\mu_i)$$

with suitable restrictions on the parameters.

6.16. With $G(s) = s^2/((s+0.01)^2 + 0.81)((s+0.01)^2 + 1.21)$, find the Hurwitz stability sector for the linear time invariant system (2.15). The nonlinear time invariant system (2.24) with $\varphi(\sigma) = 100\,\sigma^3$, on simulation using the analog computer [F 7], has been found to develop oscillations. How can one use/modify the results of the Chapter so as to confirm this analytically.

CHAPTER 7

L_2-INSTABILITY OF FEEDBACK SYSTEMS IN INTEGRAL FORM

7.1. Introduction.. We have found in Chapter 3 that a stability analysis of the feedback system

$$p(D)y + k(t)\varphi(q(D)y) = 0 \quad \text{on } [0,\infty) \tag{2.1}$$

(where k(t) and $\varphi(\cdot)$ obey Assumptions 2.21 and 2.22, p 23) amounts to establishing, under suitable assumptions on the linear time invariant system (2.15), a differential inequality of the form

$$(dV(\underline{x},t)/dt)\Big|_{(2.1)} \leqslant \lambda(t)V(\underline{x}(t),t) \tag{7.1}$$

for the solutions of (2.1), where $V(\underline{x},t)$ is a suitably chosen positive definite quadratic form in $\underline{x}$ + nonnegative functional involving $\varphi(\cdot)$. An instability analysis of (2.1), carried out in Chapter 6, involves assumptions which are counterpart to those of Chapter 3, the inequality notation in the differential inequality (7.1) being reversed.

Concerning an L_2-stability analysis of the feedback system

$$\left.\begin{aligned} v(t) &= f(t) - k(t)\varphi(\sigma(t)) \\ \sigma(t) &= \sum_{i=1} g_i v(t-\tau_i) + \int_0^{\infty} g(\tau)v(t-\tau)d\tau \end{aligned}\right\} \text{on } [0,\infty) \tag{2.7}$$

(where k(t) and $\varphi(\cdot)$ obey the same assumptions as for the system (2.1), and the assumptions on g_i and $g(\cdot)$ are specified on p 26 and p 31), an L_2-norm inequality of the type

$$\left.\begin{aligned} \|v\|^2 &\leqslant \text{const.}\ \|f\| \cdot \|v\| + \text{const.} \\ \text{or} \quad \|\sigma\|^2 &\leqslant \text{const.}\ \|f\| \cdot \|\sigma\| + \text{const.} \end{aligned}\right\} \tag{7.2}$$

was established for the solutions of (2.7). The Cauchy-Schwarz inequality (Lemma 5.33) and the Parseval relation (Lemma 4.41) formed the basis for (7.2).

In the present chapter, the converse (or counterpart), for instance Lemma 7.21, of the standard Cauchy-Schwarz inequality is used to derive in Sec.7.2 (under Assumption U2, Sec.2.32, p 32) for the solutions of (2.7), an L_2-norm inequality of the form

$$\left.\begin{aligned} &\int_0^{\infty} v^2(t)\exp(-2\varepsilon t)dt \geqslant \eta \, \|v\| \cdot \|f\| \\ \text{or} \quad &\int_0^{\infty} \sigma^2(t)\exp(-2\varepsilon t)dt \geqslant \eta \, \|\sigma\| \cdot \|f\| \end{aligned}\right\} \tag{7.3}$$

for some positive constants ε and η, for at least some inputs $f \in L_2(0,\infty)$. That is, we deal with Problem 2.42 (Sec.2.32, p 33) and the counterparts of the results in Chapter 5. Observe that inequality (7.3) implies the existence of inputs $f(\cdot) \in L_2(0,\infty)$ which give rise to $\sigma(\cdot)$, $v(\cdot)$ not in $L_2(0,\infty)$.

A review of the literature is made in Sec.7.3. The conclusions (Sec.7.4) point up the need for a resolution of basic instability problems.

7.2. Derivation of Instability Conditions.

We refer the reader to pp 28-29 and p 31 for the notation and definitions of terms below.

Corresponding to the Basic Result 6.51 for the system (2.1) which defines the shift factor β, we have the following parallel result.

Basic Result.7.21. If Assumption U2 holds, then the inequality

$$\inf_{\text{Re } s \geqslant 0} \left| 1 + KG_1(s+\beta) \right| > 0 \tag{7.4}$$

for all K in $[0,\infty)$ is not satisfied for any $\beta \leqslant 0$ but holds for some $\beta \geqslant \beta_i > 0$. β_i is the minimum value of β for which (7.4) holds for all K in $[0,\infty)$; when there is no finite value of β, we set $\beta_i = \infty$.

Remark 7.21. Satisfaction of the inequality (7.4) for any other range of variation of K can be transformed to the range $[0,\infty)$. See Exercise 2.2.

7.21. The main idea used in instability analysis.

It is well known [B 5] that no inequality of the form

$$\left(\int_0^{\infty} u_1(t)u_2(t)dt\right)^2 \geqslant \eta \left(\int_0^{\infty} u_1^2(t)dt\right)^2 \left(\int_0^{\infty} u_2^2(t)dt\right) \tag{7.5}$$

can hold for all u_1 and u_2 in $L_2(0,\infty)$ with a positive constant η. However, inequality (7.5) can hold if u_1 and u_2 can belong to certain subspaces of $L_2(0,\infty)$. See Bellman [B 4(b)] for a specific example of this type. Another result is due to Moore [M 12] :

Lemma 7.21. Let $u_1(\cdot)$ and $u_2(\cdot)$ be any nonzero functions in $L_2(0,\infty)$. Then for any nonzero function $u_3(\cdot)$ in $L_2(0,\infty)$

$$\left| \int_0^\infty u_1(t)u_2(t)dt \right| \geqslant \gamma_{u_3} \|u_1\| \cdot \|u_2\| \tag{7.6}$$

where

$$\gamma_{u_3} = \max\left\{4 \min \left(\frac{\left|\int_0^\infty u_1(t)u_3(t)dt\right|}{\|u_1\| \cdot \|u_3\|}, \frac{\left|\int_0^\infty u_2(t)u_3(t)dt\right|}{\|u_2\| \cdot \|u_3\|}\right) - 3, 0\right\}$$

Remark 7.22. Moore [M 12] establishes a converse of the Cauchy-Schwarz inequality for a real inner product space. If the inequality (7.6) is to be nontrivial, we require that $\gamma_{u_3} \neq 0$. Consequently, u_1, u_2 and u_3 are to be suitably chosen. This implies restriction on u_1, u_2 and u_3 to belong to a subspace of $L_2(0,\infty)$. However, what is of interest to us is the fact that a nontrivial subspace of $L_2(0,\infty)$ exists in which inequality (7.5) holds for a positive constant η. See also, in this context, Bellman and Beckenbach [B 5, pp 39-45] .

7.22. Preliminaries.

For any real valued function $x(\cdot)$ on $[0,\infty)$ and any $T > 0$, let $x_T(\cdot)$ denote the truncated function defined by

$$x_T(\cdot) \overset{=}{=} \begin{cases} x(t) & \text{for } t \leqslant T \\ 0 & \text{for } t > T . \end{cases}$$

Let $L_{2e}(0,\infty)$ be the space of those real valued functions $x(\cdot)$ on $[0,\infty)$ whose truncations $x_T(\cdot)$ belong to $L_2(0,\infty)$ for all $T > 0$.

In addition to the assumptions on $k(t)$ and $\varphi(\cdot)$ (Sec.2.2, p 23) and Assumption U2 (Sec.2.32, p 32) needed in the derivation of instability criteria, we make the following assumption.

Assumption 7.21. $f(\cdot)$ is in $L_2(0,\infty)$ and $v(\cdot)$, $\sigma(\cdot)$ are in $L_{2e}(0,\infty)$. This assumption is needed to guarantee applicability of the Parseval relation (Lemma 4.41) in the proof of the instability results. However, see Remark 7.24 below .

As before, let $((dk/dt)/k(t)) = \theta(t) = \theta^+(t) + \theta^-(t)$, where $\theta^+(t)$ represents the nonnegative lobes of $\theta(t)$, $\theta^-(t)$ the negative lobes of $\theta(t)$. Further

$$\delta_s = \sup_{\substack{\sigma \\ \sigma\neq 0}} \left\{ \int_0^{\sigma} \varphi(w)dw/\varphi(\sigma)\sigma \right\},$$

$$\delta_m = \inf_{\substack{\sigma \\ \sigma\neq 0}} \left\{ \int_0^{\sigma} \varphi(w)dw/\varphi(\sigma)\sigma \right\}. \tag{7.7}$$

Let $\mathcal{L}$ denote the class of operators $Z : L_{2e} \rightarrow L_{2e}$. An operator Z is causal if for all t

$$(Zx_1)(\tau) = (Zx_2)(\tau) \quad \text{for } \tau \leq t$$

whenever $x_1(\tau) = x_2(\tau)$ for $\tau \leq t$, and $x_1, x_2 \in L_{2e}$. A causal operator $Z_c \in \mathcal{L}$ of interest to us is the convolution operator of the form

$$(Z_c x)(t) = \int_0^{\infty} z_c(\tau)x(t-\tau)d\tau$$

where $z_c(t) = 0$ for $t < 0$ and

$$\int_0^{\infty} |z_c(t)| dt < \infty.$$

An anticausal operator $Z_{ac} \in \mathcal{L}$ of interest to us is the convolution operator of the form

$$(Z_{ac}x)(t) = \int_{-\infty}^{0} z_{ac}(\tau)x(t-\tau)d\tau \tag{7.8}$$

where $z_{ac}(t) = 0$ for $t > 0$ and

$$\int_{-\infty}^{0} |z_{ac}(t)| dt < \infty .$$

A (memoryless + anticausal) operator $Z_a \in \mathcal{L}$ is of the form

$$(Z_a x)(t) = \int_{-\infty}^{0} z_{ac}(\tau)x(t-\tau)d\tau + x(t)$$

where $z_{ac}(t) = 0$ for $t > 0$, and

$$\int_{-\infty}^{0} |z_{ac}(t)| dt < \infty .$$

The corresponding frequency functions are given by

$$Z_c(j\omega) = \int_0^{\infty} z_c(t)\exp(-j\omega t)dt \quad ;$$

$$Z_{ac}(j\omega) = \int_{-\infty}^{0} z_{ac}(t)\exp(-j\omega t)dt \quad ; \text{ and}$$

$$Z_a(j\omega) = 1 + \int_{-\infty}^{0} z_{ac}(t)\exp(-j\omega t)dt \ .$$

See Lemma 3.22 (p 57) concerning the existence of a noncausal multiplier function for the linear time invariant system (2.15) when it is asymptotically stable. As for the corresponding linear time invariant system

$$\begin{aligned} v(t) &= f(t) - K\,\sigma(t) \\ \sigma(t) &= \sum_{i=1} g_i v(t-\tau_i) + \int_0^{\infty} g(\tau)v(t-\tau)d\tau , \end{aligned} \qquad (2.17)$$

such a multiplier function form of its L_2-stability condition (2.19) is not known and hence is left as an exercise (Exercise 7.2).

The class $\mathcal{K}$ of time-multiplier functions $h(\cdot)$ is given in Definition 3.41, p 69 .

7.22. Main Results. Linear System L_2-Instability.

Based on the above preliminaries, we deal with the nonlinear system (2.7) and the linear system

$$\left.\begin{aligned} v(t) &= f(t) - k(t)\sigma(t) \\ \sigma(t) &= \sum_{i=1} g_i v(t-\tau_i) + \int_0^{\infty} g(\tau)v(t-\tau)d\tau \end{aligned}\right\} \text{ on } [0,\infty) \qquad (2.23)$$

separately.

As in the proof of the L_2- stability criteria in Chapter 5, the second mean value theorem (Lemma 4.31, p 99) and the Parseval relation (Lemma 4.41, p 111) play an important role in the proof of L_2-instability criteria. In Chapter 5, it was found that the causal part of the multiplier function gives rise to an upper global bound on $\theta^+(t)$ for L_2-stability of (2.7), and the anticausal part to a lower global bound on $\theta^-(t)$. Here, it turns out that an anticausal multiplier function gives rise to an upper global bound on

$\theta^+(t)$ for L_2-instability of (2.7).

We first present an instability criterion for the linear system (2.23).

Theorem 7.21. The linear system (2.23) under Assumption U2 (Sec.2.32, p 32) has solutions with property (7.3) (for $f(\cdot)$ belonging to a nontrivial subspace of $L_2(0,\infty)$) if there exists an anticausal multiplier function $Z_{ac}(j\omega)$ such that

(i) $\quad \mathrm{Re}\, Z_{ac}(j\omega)G_1(j\omega+\beta) \geq \delta > 0 \qquad (7.9)$

for some constants δ, $\beta > \beta_i$ (the latter obtained from Basic Result 7.21), for all real ω ;

(ii) $\quad \sup_{\omega} \left| Z_{ac}(j\omega)G_1(j\omega+\beta) \right| < \infty$,

$$\sup_{\omega} \left| Z_{ac}(j\omega-\beta) \right| < \infty \qquad (7.10)$$

for the value of β obtained from hypothesis (i) ;

(iii) $\quad \mathrm{Re}\, Z_{ac}(j\omega) \geq 0$ for all real ω ; $\qquad (7.11)$

and (iv) with $h(t) \in \mathcal{K}$, $h(t)k(t)\exp(-2\mu t)$ is nonincreasing for some constant $\mu < \beta$ obtained from hypothesis (i).

Proof. The proof involves four steps :

Step 1. Choice of a suitable quadratic functional $\rho(T)$. Derivation of conditions for $\rho(T)$ to have the property

$$\eta_1 \| \sigma \|^2 \leq \rho(T) \leq \eta_2 \| \sigma_\mu \|^2 \qquad (7.12)$$

for some positive constants η_1 and η_2 (which are independent of T), where $\sigma_\mu(t)$ $\exp(-\mu t)\sigma(t)$ and $\mu > 0$.

Step 2. Relating the positivity conditions of Step 1 to the frequency domain conditions given as hypotheses (i) and (ii)..

Step 3. Evaluation of $\rho(T)$ along the trajectories of (2.23). Establishing a lower bound on $\rho(T)$ so evaluated :

$$\rho(T) \Big|_{(2.23)} \geq \eta_3 \| \sigma \| \cdot \| f \| \qquad (7.13)$$

for some positive constant η_3 independent of T.

Step 4. Combining inequalities (7.12) and (7.13) to conclude property (7.3) for the solutions of (2.23).

Let

$$(\mathcal{G}x)(t) \triangleq \sum_i g_i x(t-\tau_i) + \int_0^\infty g(\tau)x(t-\tau)d\tau, \tag{7.14}$$

$$\begin{aligned}&\sigma_1(t) \triangleq k(t)\sigma(t);\ \sigma_\beta(t) = \exp(-\beta t)\sigma(t),\ \sigma_{1\beta}(t) = \exp(-\beta t)\sigma_1(t);\\ &\sigma^\varepsilon(t) = \exp(\varepsilon t)\sigma(t),\ \sigma_1{}^\varepsilon(t) = \exp(\varepsilon t)\sigma_1(t);\ \text{and}\ \sigma_{1\beta}{}^\varepsilon(t) = \exp(\varepsilon t)\sigma_{1\beta}(t)\end{aligned} \tag{7.15}$$

We have, from (2.23), (7.14) and (7.15)

$$\sigma(t) = (\mathcal{G}f)(t)-(\mathcal{G}\sigma_1)(t) \tag{7.16}$$

and hence

$$\sigma_\beta(t) = \exp(-\beta t)(\mathcal{G}f)(t) - (\mathcal{G}_\beta\sigma_{1\beta})(t) \tag{7.17}$$

where

$$(\mathcal{G}_\beta x_\beta)(t) = \exp(-\beta t)(\mathcal{G}x)(t) \tag{7.18}$$

Further, let $x_{\beta T}$ denote the truncated function : $x_{\beta T} = x_\beta$ for $0 \leq t \leq T$
$= 0$ for $t > T$.

Steps 1 and 2. Consider the functional

$$\rho(T) = \int_0^T h(t)k(t)\sigma_\beta(t)(\mathcal{Z}_{ac}\mathcal{G}_\beta\sigma_{1\beta})(t)dt + \int_0^T h(t)k(t)\sigma_\beta(t)(\mathcal{Z}_{ac}\mathcal{G}_\beta\sigma_{1\beta})(t)dt \tag{7.19}$$

where $h(\cdot)\in\mathcal{K}$, and $\mathcal{Z}_{ac}\in\mathcal{L}$ is an anticausal convolution operator of the form given by (7.8).

Since $h(\cdot)\in\mathcal{K}$, $h(t)\exp(-2\varepsilon t)$ is nonincreasing for some $\varepsilon > 0$. The second integral on the right hand side of (7.19) can be written as

$$\int_0^T h(t)k(t)\sigma_\beta(t)(\mathcal{Z}_{ac}\mathcal{G}_\beta\sigma_{1\beta})(t)dt = \int_0^T h(t)\exp(-2\varepsilon t)\sigma_{1\beta}(t)\exp(2\varepsilon t)(\mathcal{Z}_{ac}\mathcal{G}_\beta\sigma_{1\beta})(t)dt$$

$$= \int_0^T h(t)\exp(-2\varepsilon t)\sigma_{1\beta}^\varepsilon(t)(\mathcal{Z}_{ac}^\varepsilon\mathcal{G}_\beta^\varepsilon\sigma_{1\beta}^\varepsilon)(t)dt \tag{7.20}$$

where

$$\begin{aligned}(\mathcal{Z}_{ac}^\varepsilon x^\varepsilon)(t) &= \exp(\varepsilon t)(\mathcal{Z}_{ac}x)(t)\\ (\mathcal{G}_\beta^\varepsilon x^\varepsilon)(t) &= \exp(\varepsilon t)(\mathcal{G}_\beta x)(t).\end{aligned} \tag{7.21}$$

By virtue of the second mean value theorem (Lemma 4.31, p 99), there is a point T' in $[0,T]$ such that the right hand side of (7.20) becomes

$$h(0)\int_0^{T'} \sigma_{1\beta}^{\varepsilon}(t)(\mathcal{Z}_{ac}^{\varepsilon}\, \mathcal{G}_{\beta}^{\varepsilon}\, \sigma_{1\beta}^{\varepsilon})(t)dt.$$

Hence, from (7.20) we have

$$\int_0^T h(t)\sigma_{1\beta}(t)(\mathcal{Z}_{ac}\mathcal{G}_{\beta}\sigma_{1\beta})(t)dt = h(0)\int_0^{T'} \sigma_{1\beta}^{\varepsilon}(t)(\mathcal{Z}_{ac}^{\varepsilon}\,\mathcal{G}_{\beta}^{\varepsilon}\,\sigma_{1\beta}^{\varepsilon})(t)dt \tag{7.22}$$

Similarly, the first integral on the right hand side of (7.19) can be written as

$$\int_0^T h(t)k(t)\sigma_{\beta}(t)(\mathcal{Z}_{ac}\sigma_{\beta})(t)dt = \int_0^T h(t)k(t)\exp(-2\mu t)\sigma_{\beta}(t)\exp(2\mu t)(\mathcal{Z}_{ac}\sigma_{\beta})(t)dt$$

$$= \int_0^T h(t)k(t)\exp(-2\mu t)\sigma_{\beta}^{\mu}(t)(\mathcal{Z}_{ac}^{\mu}\,\sigma_{\beta}^{\mu})(t)dt \tag{7.23}$$

By virtue of hypothesis (iv), $h(t)k(t)\exp(-2\mu t)$ is nonincreasing for some $\mu > 0$. Invoke the second mean value theorem (Lemma 4.31, p 99) to conclude from (7.23) that there is a point T'' in $[0,T]$ such that

$$\int_0^T h(t)k(t)\sigma_{\beta}(t)(\mathcal{Z}_{ac}\,\sigma_{\beta})(t) = h(0)k(0)\int_0^{T''} \sigma_{\beta}^{\mu}(t)(\mathcal{Z}_{ac}^{\mu}\,\sigma_{\beta}^{\mu})(t)dt \tag{7.24}$$

Hence combining (7.22) and (7.24) in (7.19), we get, by virtue of hypothesis (iv),

$$\rho(T) = h(0)k(0)\int_0^{T''} \sigma_{\beta}^{\mu}(t)(\mathcal{Z}_{ac}^{\mu}\,\sigma_{\beta}^{\mu})(t)dt + h(0)\int_0^{T'} \sigma_{1\beta}^{\varepsilon}(t)(\mathcal{Z}_{ac}^{\varepsilon}\mathcal{G}_{\beta}^{\varepsilon}\sigma_{1\beta}^{\varepsilon})(t)dt \tag{7.25}$$

The first integral, I_1, on the right hand side of (7.25) involves the anticausal operator $\mathcal{Z}_{ac}^{\mu}$ defined by

$$(\mathcal{Z}_{ac}^{\mu}\, x)(t) = \int_{-\infty}^{0} z_{ac}(\tau)\exp(\mu\tau)x(t-\tau)d\tau \; .$$

Hence, by virtue of hypothesis (iii) and the fact that μ is chosen to be positive, the Parseval relation (Lemma 4.41, p 111) yields the result that I_1 is nonnegative and has the property

$$0 \leqslant I_1 \leqslant h(0)k(0)\sup_{\omega}\left|Z_{ac}(j\omega-\mu)\right|\,\|\sigma_{\beta}^{\mu}\|^2 = \eta_1\,\|\sigma_{\beta T}^{\mu}\|^2 \tag{7.26}$$

where $\eta_1 = h(0)k(0) \sup_{\omega} \left| Z_{ac}(j\omega-\mu) \right| < \infty$ from hypothesis (ii).

The second integral, I_2, on the right hand side of (7.25) by a suitable choice of $\beta \geq \beta_i + \varepsilon$ and by virtue of hypothesis (i), can be shown, by invoking the Parseval relation, to be positive and to have the property

$$\eta_2 \left\| \sigma_{1\beta T}^{\varepsilon} \right\|^2 \leq I_2 \leq \eta_3 \left\| \sigma_{1\beta T}^{\varepsilon} \right\|^2 \tag{7.27}$$

where $\eta_2 > 0$ and $\eta_3 = h(0) \sup_{\omega} \left| Z_{ac}(j\omega-\varepsilon)G_1(j\omega+\beta-\varepsilon) \right| < \infty$ from hypothesis (ii)..Combining the bounds (7.26) and (7.27) on I_1 and I_2 respectively and substituting into (7.25), we get

$$\eta_2 \left\| \sigma_{1\beta T}^{\varepsilon} \right\|^2 \leq \rho(T) \leq \eta_1 \left\| \sigma_{\beta T}^{\mu} \right\|^2 + \eta_3 \left\| \sigma_{1\beta T}^{\varepsilon} \right\|^2 \tag{7.28}$$

Step 3. The functional $\rho(T)$, defined by (7.19), along the trajectories of (2.23) assumes the value

$$\rho(T) \Big|_{(2.23)} = \int_0^T h(t)k(t)\sigma_{1\beta}(t)(Z_{ac}\mathcal{G}_\beta f_\beta)(t)dt$$

$$= \int_0^T h(t)k(t)\exp(-2\beta t)\sigma_{1\beta}(t)\exp(2\beta t)(Z_{ac}\,\mathcal{G}_\beta\, f_\beta)(t)dt. \tag{7.29}$$

By virtue of hypothesis (iv), an application of the second mean value theorem to (7.29) yields the result that there is a point T_1 in $[0,T]$ such that

$$\rho(T) \Big|_{(2.23)} = h(0)k(0) \int_0^{T_1} \sigma_{1\beta}(t)\exp(2\beta t)(Z_{ac}\mathcal{G}_\beta f_\beta)(t)dt. \tag{7.30}$$

Recalling the definitions (7.15) and (7.18), (7.30 can be rewritten as

$$\rho(T) \Big|_{(2.23)} = h(0)k(0) \int_0^{T_1} \sigma(t)(Z_{ac}^{\beta}\,\mathcal{G}f)(t)dt \tag{7.31}$$

Note that no assumption of boundedness on $\mathcal{G}$ has been made.

Invoking the Parseval relation and the converse Cauchy-Schwarz inequality, we find that $\rho(T)\big|_{(2.23)}$ has, in some nontrivial subspace of $L_2(0,\infty)$, the property

$$\rho(T)\Big|_{(2.23)} \geq h(0)k(0)\,\alpha \left\|\sigma_{T_1}\right\| \cdot \left\|f_{T_1}\right\| \tag{7.32}$$

for some α.

But $\rho(T)$ cannot be nonpositive in view of the property (7.28). Hence, from (7.32)

$$\rho(T)\Big|_{(2.23)} \geq \eta_4 \|\sigma_{T_1}\| \cdot \|f_{T_1}\| \tag{7.33}$$

for some constant $\eta_4 > 0$, for f_{T_1} belonging to some nontrivial subspace of $L_2(0,\infty)$.

Step 4. Combine inequalities (7.28) and (7.33) to conclude that

$$\eta_1 \|\sigma^{\mu}_{\beta T}\|^2 + \eta_3 \|\sigma^{\varepsilon}_{1\beta T}\|^2 \geq \eta_4 \|\sigma_{T_1}\| \cdot \|f_{T_1}\| \tag{7.34}$$

for some positive constants η_1, η_3 and η_4, for f_{T_1} belonging to some nontrivial subspace of $L_2(0,\infty)$. Since T_1 is a point in $[0, T]$, the inequality (7.34) is valid for all $T > 0$ and, by virtue of hypothesis (iv), $\mu < \beta$, the solutions of (2.23) have property (7.3) for f belonging to some nontrivial subspace of $L_2(0,\infty)$.

Note that the inequality (7.34) holds for all $\beta > \beta_i$ obtained from Basic Result 7.21. Consequently, in hypotheses (i)-(iii), choice of $\beta = \beta_i + \varepsilon$ for some $\varepsilon > 0$, and in hypothesis (iv), choice of μ β_i guarantees that the hypotheses (i)-(iv) hold for all $\beta \geq \beta_i$. The theorem is proved.

Remark 7.23. Hypothesis (iv) of Theorem 7.21 can be expressed as a global upper bound on $\theta^+(t)$. See Lemma 3.42 (p 69). Observe that Theorem 7.21, with hypothesis (iv) replaced by the global upper bound on $\theta^+(t)$, is the counterpart of Theorem 3.42 (see pp 69-71). The geometric interpretation of the instability theorem is suggested as an exercise (Exercise 7.4.).

7.23. Main Results. Nonlinear System L_2-Instability.

As regards the nonlinear system (2.7), we consider the two cases $\varphi(\cdot) \in C$ and $\varphi(\cdot) \in C_m$. The corresponding multiplier functions we choose are $(a_o + b_o j\omega)$ and $1 + Z_{ac}(j\omega)$. In fact, no (instability) multiplier function more general than $(a_o + b_o j\omega)$ is known which could be used in the instability analysis of (2.7) $\varphi(\cdot) \in C$.

The main lemmas needed in the proof of the instability theorems for (2.7) are the following :

Lemma 7.22. Given that $\varphi(\cdot)\in C$, $h(\cdot)\in\mathcal{K}$, $\beta>0$, $a_o>0$ and $b_o\geq 0$, then the functional

$$I_{n1} = \int_0^T h(t)k(t)\exp(-\beta t)\varphi(\sigma(t))(a_o\sigma_\beta(t) + b_o(d\sigma_\beta/dt))dt + b_o h(0)k(0)\Phi(\sigma(0)) \qquad (7.35)$$

is nonnegative for all $\sigma(\cdot)$, for all $T>0$ if

(i) $h(t)k(t)\exp(-2\beta t)$ is nonincreasing ; and

(ii) $\beta b_o \leq a_o$.

Proof. See Appendix 7.1.

Lemma 7.23. Given $\varphi(\cdot)$, $h(\cdot)$, β, a_o and b_o as in Lemma 7.22, then the functional

$$I_{n2} = \int_0^T h(t)k(t)\exp(-\beta t)\varphi(\sigma(t))((a_o+b_o(d/dt))G_\beta f_\beta)(t)dt \qquad (7.36)$$

satisfies the inequality

$$I_{n2}\geq \eta\ \|\sigma_{T'}\|\cdot\|f_{T'}\|$$

for some constant $\eta>0$, T' in $[0,T]$ and some $f(\cdot)$ belonging to a nontrivial subspace of $L_2(0,\infty)$ if

(i) hypothesis (i) of Lemma 7.22 holds ; and

(ii) it is known from a different source that I_{n2} is positive for all $\sigma_T(\cdot)\in L_2(0,\infty)$ and $f_T(\cdot)\in L_2(0,\infty)$, for instance, satisfying the system equation (2.7).

Proof. See Appendix 7.2.

Lemmas 7.22 and 7.23 are needed in the proof of an instability result (Theorem 7.22) for (2.7) with $\varphi(\cdot)\in C$. We now state the lemmas concerning $\varphi(\cdot)\in C_m$.

Lemma 7.24. Given that $\varphi(\cdot)\in C_m$, $h(\cdot)\in\mathcal{K}$, $\beta>0$ and the operator $Z_{ac}\in\mathcal{L}$, then the functional

$$I_{n3} = \int_0^T h(t)k(t)\exp(-\beta t)\varphi(\sigma(t))(\sigma_\beta(t) +(Z_{ac}\sigma_\beta)(t))dt \qquad (7.37)$$

is nonnegative for all $\sigma(\cdot)$ in the domain of Z_{ac}, for all $T>0$ if

(i) $z_{ac}(t) \leq 0$ for all $t \leq 0$;

(ii) for some positive constant $\mu < \beta$,

$$\int_{-\infty}^{0} |z_{ac}(t)| \exp(2\mu-\beta)t \, dt \leq 1/(1+\delta_s-\delta_m) \tag{7.38}$$

where δ_s and δ_m are defined by (7.7) ; and

(iii) $h(t)k(t)\exp(-2\mu t)$ is nonincreasing for the value of μ obtained from hypothesis (ii).

Proof. See Appendix 7.3.

Lemma 7.25. Given that $\varphi(\cdot) \in C_m$, $h(\cdot) \in \mathcal{K}$, $\beta > 0$ and the operator $Z_{ac} \in \mathcal{L}$, then the functional

$$I_{n4} = \int_0^T h(t)k(t)\exp(-\beta t)\varphi(\sigma(t))((G_\beta f_\beta)(t) + (Z_{ac}G_\beta f_\beta)(t))dt \tag{7.39}$$

satisfies the inequality

$$I_{n4} \geq \eta \, \|\sigma_{T'}\| \cdot \|f_{T'}\|$$

for some constant $\eta > 0$, T' in $[0,T]$ and some $f(\cdot)$ belonging to a nontrivial subspace of $L_2(0,\infty)$ if

(i) $h(t)k(t)\exp(-2\mu t)$ is nonincreasing for some $\mu < \beta$; and

(ii) it is known from some other source that I_{n4} is positive for all $\sigma_T(\cdot) \in L_2(0,\infty)$ and $f(\cdot) \in L_2(0,\infty)$, for instance, satisfying the system equation (2.7).

Proof. See Appendix 7.4.

We now state two main instability results for (2.7), the first concerning $\varphi(\cdot) \in C$, and the second concerning $\varphi(\cdot) \in C_m$. The proofs, based on Lemmas 7.22-7.25, are similar to the proof of Theorem 7.21 and hence are left as an exercise. (Exercise 7.5).

Theorem 7.22. The nonlinear system (2.7) with $\varphi(\cdot) \in C$ under Assumption U2 (sec. 2.32, p 32) has solutions with property (7.3) (for $f(\cdot)$ belonging to some nontrivial subspace of $L_2(0,\infty)$) if there exists a multiplier function $(a_o + b_o j\omega)$ with $a_o > 0$, $b_o \geq 0$ such that

(i) $\mathrm{Re}(a_o + b_o j\omega)G_1(j\omega+\beta) \geq \delta > 0$

for some constants δ, $\beta > \beta_i$ (the latter obtained from Basic Result 7.21), for all real ω ;

(ii) $\beta b_o < a_o$;

(iii) $\sup_{\omega} \left| (a_o + b_o j\omega) G_1(j\omega+\beta) \right| < \infty$

for the value of β obtained from hypothesis (i) ;

(iv) with $h(\cdot) \in \mathcal{K}$, $h(t)k(t)\exp(-2\beta t)$ is nonincreasing for the value of β obtained from hypothesis (i).

Theorem 7.23. The nonlinear system (2.7) with $\varphi(\cdot) \in C_m$ under Assumption U2 (Sec.2.32, p 32) has solutions with property (7.3) for $f(\cdot)$ in some nontrivial subspace of $L_2(0,\infty)$ if there exists a (memoryless + anticausal) multiplier function $(1 + Z_{ac}(j\omega))$ such that

(i) $\mathrm{Re}(1+Z_{ac}(j\omega))G_1(j\omega+\beta) \geq \delta > 0$

for some constants δ, $\beta > \beta_i$ (the latter obtained from Basic Result 7.21), for all real ω ;

(ii) $\sup_{\omega} \left| (1+Z_{ac}(j\omega))G_1(j\omega+\beta \right| < \infty$

for the value of β obtained from hypothesis (i) ;

(iii) $z_{ac}(t) \leq 0$ for all $t \leq 0$ and there exists a positive constant $\mu < \beta$ (the latter obtained from hypothesis (i)) for which the integral inequality (7.38) holds ;

(iv) with $h(\cdot) \in \mathcal{K}$, $h(t)k(t)\exp(-2\mu t)$ is nonincreasing for the value of μ obtained from hypothesis (iii).

Remark 7.24. On comparing Theorem 7.23 concerning L_2-instability of (2.7) with Theorem 6.54 (p 184) concerning exponential instability of (2.1) we note the following points of difference between the two : (i) The multiplier function of the former is (memoryless + anticausal) but that of the latter is (memoryless + causal) ;

(ii) The upper bound on the time-domain integral constraint on $z_{ac}(t)$ of the former is $1/(1+\delta_s-\delta_m)$ but that on $z_g(t)$ of the latter is $1/(2+\delta_s-\delta_m)$;

(iii) The former involves an upper bound on $\theta^+(t)$ but the latter involves, in essence, a lower bound on $\theta^-(t)$.

However it is not known how to take account of the discrepancy brought out in item (ii) above (Exercise 7.8).

7.24 Main Results. Instability of Operator Feedback Systems.

The system under consideration (Fig.5.1) is described by

$$\begin{aligned} v(\cdot) &= f(\cdot) - \mathcal{G}_2\sigma(\cdot) \\ \sigma(\cdot) &= \mathcal{G}_1 v(\cdot) \end{aligned} \tag{7.40}$$

where $\mathcal{G}_1$ and $\mathcal{G}_2$ are general causal operators. The lemmas and theorems of Sec. 7.22 and Sec.7.23 form the basis for an operator-theoretic statement of L_2-instability results concerning the system (7.40). For definitions,and notation on operator manipulations, see Sec.5.2 (pp 142-143) and Sec.5.4 (pp 150-151).

Once again, we make the assumption (Assumption 7.21) that the response σ and the error signal v satisfy the conditions $\sigma(\cdot) \in L_{2e}$ and $v(\cdot) \in L_{2e}$ for each input $f(\cdot) \in L_2(0,\infty)$.

Remark 7.25 Any instability (or stability) result with a claim to generality should dispense with the above assumption of 'no finite escape time'. In fact, the instability result by itself should include the possibility of finite escape time for the system trajectories. (Analogously, the stability result by itself should preclude the possibility of finite escape time for the system trajectories). However, the present framework does not permit derivation of such an instability criterion. The assumption of $\sigma(\cdot) \in L_{2e}$ and $v(\cdot) \in L_{2e}$ for each input $f(\cdot) \in L_2(0,\infty)$ is needed to guarantee existence of the functional used in deriving the instability criterion.

Let, as before,

$$x_\beta(t) \triangleq \exp(-\beta t)x(t) \ ; \qquad x^\beta(t) = \exp(\beta t)x(t) \ ;$$

$$(\mathcal{G}'_{1\beta}x)(t) = \exp(-\beta t)(\mathcal{G}_1 x^\beta)(t) \ ; \quad (\mathcal{G}'_{2\beta}x)(t) = \exp(-\beta t)(\mathcal{G}_2 x^\beta)(t) \ ;$$

and $x_{\beta T}(\cdot)$, $x_T^\beta(\cdot)$,$(G'_{1\beta}x_T)(\cdot)$, and $(G'_{2\beta}x_T)(\cdot)$ denote the corresponding truncated versions..

We now prove the following instability counterpart of the passivity stability theorem (Theorem 5.41, p 151) [Z 3(b)] of Chapter 5. The instability theorem unifies the earlier results of Sec.7.22 and Sec.7.23, derived for the convolution operator feedback system (2.7).

Theorem 7.24. The feedback system (7.40) has solutions with property (7.3) for some $f(\cdot)$ belonging to a nontrivial subspace of $L_2(0,\infty)$ if there exists an operator $\mathcal{Z} \in$ class $\mathcal{L}$ of operators mapping $L_{2e} \rightarrow L_{2e}$ with its adjoint $\mathcal{Z}^* \in \mathcal{L}$ such that

(i) $\mathcal{Z}\, \mathcal{G}'_{1\beta}$, $\mathcal{Z}^* \mathcal{G}'_{2\beta}$ are passive and have finite gains for some $\beta > 0$;

either (iia) $\mathcal{Z}^* \mathcal{G}'_{2\beta}$ is strictly passive, $\mathcal{G}_2$ has a finite gain or

(iib) $\mathcal{Z}\, \mathcal{G}'_{1\beta}$ is strictly passive, $\mathcal{G}_1$ has a finite gain ; and

(iii) for no nonpositive value of β, hypothesis (i) is satisfied.

Proof. As in the proof of Theorem 7.21, the proof consists of four steps :

Step 1. Choice of a suitable quadratic functional $\rho_1(T)$. Derivation of conditions for $\rho_1(T)$ to have the property

$$\eta_1 \, \|\sigma_T\|^2 \leq \rho_1(T) \leq \eta_2 \, \|\sigma_{\varepsilon T}\|^2 \tag{7.41}$$

for some positive constants ε, η_1 and η_2 independent of T.

Step 2. Relating the conditions of Step 1 to the passivity of operators involving $\mathcal{G}_1$ and $\mathcal{G}_2$.

Step 3. Evaluation of $\rho_1(T)$ along the trajectories of (7.40). Establishing a lower bound on $\rho_1(T)$ so evaluated:

$$\rho_1(T)\Big|_{(7.40)} \geq \eta_3 \, \|\sigma_T\| \cdot \|f_T\| \tag{7.42}$$

for some positive constant η_3 independent of T, for some $f(\cdot)$ belonging to a nontrivial subspace of $L_2(0,\infty)$.

Step 4. Combining inequalities (7.41) and (7.42) to conclude that the solutions of (7.40) have property (7.3).

Part I. Use of hypotheses (i), (iia) and (iii).

Let

$$\rho_1(T) = \int_0^T \exp(-\beta t)(\mathcal{G}_2\sigma)(t)(\mathcal{Z}\, \sigma_\beta)(t)dt$$
$$+ \int_0^T \exp(-\beta t)(G_2\sigma)(t)(\mathcal{Z}\, \mathcal{G}'_{1\beta}\mathcal{G}'_{2\beta}\sigma_\beta)(t)dt. \tag{7.43}$$

Steps 1 and 2. The first integral on the right hand side of (7.43) can be written as

$$I' = \int_0^T \exp(-\beta t)(\mathcal{G}_2\sigma)(t)(\mathcal{Z}\sigma_\beta)(t)dt$$

$$= \int_0^T (\mathcal{G}'_{2\beta}\sigma_\beta)(t)(\mathcal{Z}\sigma_\beta)(t)dt = \int_0^T (\mathcal{Z}^*\mathcal{G}'_{2\beta}\sigma_\beta)(t)\sigma_\beta(t)dt$$

Hence, by virtue of hypothesis (i) and (iia), we get

$$\alpha_1 \|\sigma_{\beta T}\|^2 \leq I' \leq \alpha_2 \|\sigma_{\beta T}\|^2 \tag{7.44}$$

for some positive constants α_1 and α_2 independent of T. Similarly, the second integral on the right hand side of (7.43) can be written as

$$I'' = \int_0^T \exp(-\beta t)(\mathcal{G}_2\sigma)(t)(\mathcal{Z}\,\mathcal{G}'_{1\beta}\mathcal{G}'_{2\beta}\sigma_\beta)(t)dt$$

$$= \int_0^T (\mathcal{G}'_{2\beta}\sigma_\beta)(t)(\mathcal{Z}\mathcal{G}'_{1\beta}\mathcal{G}'_{2\beta}\sigma_\beta)(t)dt$$

which, by virtue of hypotheses (i) and (iia), satisfies the inequality

$$0 \leq I'' \leq \alpha_3 \|\sigma_{\beta T}\|^2 \tag{7.45}$$

for some positive constant α_3 independent of T.

Combining the inequalities (7.44) and (7.45), we infer that hypotheses (i) and (iia) imply

$$\alpha_1 \|\sigma_{\beta T}\|^2 \leq \rho_1(T) \leq \alpha_4 \|\sigma_{\beta T}\|^2 \tag{7.46}$$

for some positive constants α_1 and α_4 (with $\alpha_1 < \alpha_4$).

Step 3. $\rho_1(T)$, as evaluated along the trajectories of (7.40), is given by

$$\rho_1(T)\Big|_{(7.40)} = \int_0^T \exp(-\beta t)(\mathcal{G}_2\sigma)(t)(\mathcal{Z}\,f_1)(t)dt \tag{7.47}$$

where

$$f_1(t) \triangleq \exp(-\beta t)(\mathcal{G}_1 f)(t).$$

Note that (i) no assumption on the boundedness of the gain of $\mathcal{G}_1$ in hypothesis (iia) is made ; and

(ii) $\rho_1(T)$ is positive for all $\sigma_T \neq 0$.

We invoke the converse Cauchy-Schwarz inequality (Sec.7.21) to conclude, from (7.47), that, in a nontrivial subspace of $L_2(0,\infty)$, there exists an $f_T(\cdot)$ and, corresponding to it, $\sigma_T(\cdot)$ as a solution of (7.40), such that

$$\rho_1(T)\Big|_{(7.40)} \geq \eta \, \|\sigma_T\| \cdot \|f_T\| \tag{7.48}$$

for some positive constant η independent of T.

Step 4. Combine inequalities (7.46) and (7.48) to get, for the solutions of (7.40),

$$\alpha_4 \, \|\sigma_{\beta T}\|^2 \geqslant \eta \, \|\sigma_T\| \cdot \|f_T\|. \tag{7.49}$$

Inequality (7.49), concerning the solutions of (7.40), has been established for all $T>0$. In view of the hypothesis (iii), we conclude that $\sigma(\cdot)$ of the system (7.40) has property (7.3) for some $f(\cdot)$ belonging to a nontrivial subspace of $L_2(0,\infty)$.

Further, from (7.40), we have

$$v(\cdot) = f(\cdot) - \mathcal{G}_2\sigma(\cdot) ,$$

from which, using the Cauchy-Schwarz inequality, we get

$$\|v\|^2 \geqslant \|f\|^2 + \|\mathcal{G}_2\sigma\|^2 - 2\|f\| \cdot \|\mathcal{G}_2\sigma\| = (\|f\| - \|\mathcal{G}_2\sigma\|)^2. \tag{7.50}$$

But $\sigma(\cdot)$ has the property (7.3). Hence $v(\cdot)$ also has the property (7.3).

Part II. Use of hypotheses (i), (iib) and (iii).

Let

$$\rho_2(T) = \int_0^T \exp(-\beta t)v(t)(\mathcal{Z}\mathcal{G}_{1\beta}' v_\beta)(t)dt + \int_0^T (\mathcal{G}_{2\beta}'\sigma_\beta)(t)(\mathcal{Z}\sigma_\beta)(t)dt .$$

Proceeding in accordance with Steps 1-4, we get, for the solutions of (7.40), the inequality

$$\alpha_1\|v_{\beta T}\|^2 \leqslant \rho_2(T) \leqslant \alpha_2\|v_{\beta T}\|^2 + \alpha_3\|\sigma_{\beta T}\|^2 \tag{7.51}$$

and

$$\rho_2(T)\Big|_{(7.40)} \geqslant \eta_1 \|\sigma_T\| \cdot \|f_T\| \tag{7.52}$$

for some positive constants α_1, α_2, α_3 and η_1, for some $f_T(\cdot)$ belonging to a nontrivial subspace of $L_2(0,\infty)$.

Hence, combining the inequalities (7.51) and (7.52), we get, for the solutions of (7.40), the inequality

$$\alpha_2 \| v_{\beta T} \|^2 + \alpha_3 \| \sigma_{\beta T} \|^2 \geq \eta_1 \| \sigma_T \| \cdot \| f_T \| \tag{7.53}$$

for some $f_T(\cdot)$ belonging to a nontrivial subspace of $L_2(0,\infty)$, for all $T > 0$. But, from (7.40),

$$\sigma(\cdot) = G_1 v(\cdot)$$

and, by virtue of hypothesis (iib), G_1 has a finite gain. Consequently, the solution inequality (7.53) becomes

$$\alpha_4 \| \sigma_{\beta T} \|^2 \geq \eta_1 \| \sigma_T \| \cdot \| f_T \| \tag{7.54}$$

for some $f_T(\cdot)$ belonging to a nontrivial subspace of $L_2(0,\infty)$, for all $T > 0$. By virtue of hypothesis (iii), $\beta > 0$. Therefore, $\sigma(\cdot)$ of the system (7.40) has property (7.3). Further, from (7.50), $v(\cdot)$ of the system (7.40) has property (7.3). The theorem is proved.

Corollary 7.241. The feedback system of Fig.5.1, described by (7.40), has solutions with property (7.3) if

(i) $\mathcal{G}'_{1\beta}, \mathcal{G}'_{2\beta}$ are passive and have finite gains for some $\beta > 0$;

either (iia) $\mathcal{G}'_{2\beta}$ is strictly passive, $\mathcal{G}_2$ has a finite gain

or (iib) $\mathcal{G}'_{1\beta}$ is strictly pasive, $\mathcal{G}_1$ has a finite gain ;

and (iii) for no nonpositive value of β, hypothesis (i) is satisfied.

Remark 7.26. The finite gain version of Corollary 7.241 gives the circle instability criterion (Exercise 7.5). Invoking the Gohberg-Krein lemma (Lemma 5.42, p 155) on the factorization of a passive operator, Theorem 7.24 can be more rigorously formulated. This improvement is left as an exercise.

7.3. Instability Criteria of Literature.

As indicated above in Sec.7.1, the basis for the stability analysis of feedback systems by functional methods á la Zames [Z 3(b)] and Sandberg [S 2] is the well known Cauchy-Schwarz inequality as applied to properly chosen functionals. In contrast with this, instability analysis found in literature has followed a different course in the hands of research workers. Here we review the instability results of some authors who employ functional methods. For a reference to the instability results obtained in the Lyapunov framework, see Chapter 6 (Sec.6.6, p 187).

For clarity, the review is organized under the following subheads as applied to the feedback system (2.7) :

(a) Assumptions (other than those used in Sec.7.2) ;

(b) Main idea in the proof of the instability criteria ; and

(c) Comments, if any, on the instability criteria of the literature.

(1) Willems [W 3(b)-(d)].

(a) Assumption. The open loop is stable. That is, $\mathcal{G}_1$ is a stable block.

(b) Main idea. Let the relation between the input $f(\cdot)$ and the output $\sigma(\cdot)$ of the linear system (2.23) be described by

$$(\mathcal{H}\sigma)(t) = f(t), \quad t \in [0,\infty).$$

where $\mathcal{H}$ is an operator on $L_2(0,\infty)$. It is to be shown that $\mathcal{H}$ is noninvertible on $L_2(0,\infty)$. To this end, extend $\mathcal{H}$ to an operator $\mathcal{H}'$ on $L_2(-\infty,\infty)$. "It can very well happen that $\mathcal{H}'$ has a noncausal inverse on $L_2(-\infty,\infty)$. ... This leads to a demonstration of the noninvertibility of $\mathcal{H}$ on $L_2(0,\infty)$." [W 3(b), p 662] .

(c) Comments. The explanation of how to demonstrate instability [W 3(b), p 647, pp 658-662] is, at least to a beginner, baffling. How does one employ all the complicated arguments in a particular problem, for instance, the Mathieu equation, to derive conditions for instability ? Skoog (see below), who uses Willems' idea, throws some light on the efficacy of the causality-noncausality-invertibility approach.

(2) Skoog [S 5] .

(a) Assumptions. (i) In (2.7), $g_i = 0$ for all $i = 2,3,\cdots,$. and $\tau_1 = 0$. $g(t) = h_1(t) + h_2(t)$ where $h_1 \in L_1(0,\infty)$ and $h_{2\beta} \in L_1(0,\infty)$ for some $\beta \geq \beta_0 > 0$. The Laplace transform of $h_2(t)$ is a rational function with a finite number of singularities and none in Re $s \leq 0$. $G_1(s)$, defined by (2.18) with $g_i = 0$ for all $i = 2,3,$ $\cdots$, and $\tau_1 = 0$, has poles $P \neq 0$ in Re $s > 0$. Further,

Re $G_1(j\omega) \geq 0$ for all real ω .

(b) Main idea. Due to Willems [W 3(b)] . The original 'unbounded' causal system is transformed to a 'bounded' noncausal system as follows:

Define $\quad \tilde{h}_2(t) = h_2(t) \quad$ for $t \leq 0$

$\qquad\qquad = 0 \qquad$ for $t > 0$.

Let

$$(\mathcal{G}_{nc}f)(t) = g_1 f(t) + \int_0^t h_1(t-\tau)f(\tau)d\tau$$

$$+ \int_{-\infty}^{0} h_2(\tau)f(t-\tau)d\tau$$

The key result concerning $\mathcal{G}_{nc}$ is the following lemma :

Lemma (Skoog [S 5, Lemma 1, p 89]). If $f \in L_2(0,\infty)$ and $\mathcal{G}_1 f \in L_2(0,\infty)$, then $\mathcal{G}_1 f = \mathcal{G}_{nc} f$ where f and $\mathcal{G}_1 f$ are viewed as elements of $L_2(-\infty,\infty)$.

(c) Comments. The key result of Skoog is faulty because it implies the relation

$$\int_0^t h_2(t-\tau)f(\tau)d\tau = \int_{-\infty}^{0} \tilde{h}_2(\tau)f(t-\tau)d\tau \ . \tag{7.55}$$

Observe that, on the left hand side of (7.55), the contribution of $f(\cdot)$ is restricted to the time interval $[0,t)$; on the right hand side, it is restricted to the time interval $[t,\infty)$. Without apriori assumptions on $f(\cdot)$, the values of $f(\cdot)$ in $[0,t)$ could be specified independent of the values of $f(\cdot)$ in $[t,\infty)$. Since (7.55) is to hold for all $f(\cdot)$ in the chosen space, $h_2(t) \equiv 0$ for all $t \geq 0$. This is trivial and, in fact, violates the assumptions made in (a) above. Hence the criteria of Skoog [S 5] amount to nothing.

(3) Takeda and Bergen [T 1(a)] .

(a) Assumptions. The feedback gain is time invariant. $\mathcal{G}_1$ is unstable. Referring to (2.7), $g_i = 0$ for all $i = 2,3, \cdots$, and $\tau_1 = 0$. $g(t) = h_1(t) + h_2(t)$ where $h_1 \in L_1(0,\infty)$ and $h_{2\beta} \in L_1(0,\infty)$ for some $\beta \geq \beta_0 > 0$. The Laplace transform of $h_2(t)$ is a strictly proper rational function of s with all poles in Re s > 0. Poles of $G_1(s)$ on the $j\omega$-axis precluded.

(b) Main idea. The space M, a proper subspace of $L_2(0,\infty)$, is the space of inputs over which $\mathcal{G}_1$ produces L_2-bounded outputs. Let $\bar{M}$ be the closure of M. The structure of a Hilbert space permits L_2 to be decomposed into $\bar{M}$ and its orthogonal complement $M^{\perp}$. $\mathcal{G}_1$ has finite gain over M.

Let the nonlinear (time invariant) block in the feedback path be represented by $\mathcal{N}$.

In one instability criterion - the so-called small-gain instability theorem -- the following gain factors, $\gamma_m(\mathcal{G}_1)$ and $\gamma(\mathcal{N})$, appear.

$$\|\mathcal{G}_1 f\| \leq \gamma_m(\mathcal{G}_1)\ \|f\| \quad \text{for all } f \in M\ ;$$
$$\|\mathcal{N} f\| \leq \gamma(\mathcal{N})\ \|f\| \quad \text{for all } f \in L_2.$$

(Note the different spaces in which the gains are defined).

In the other instability criterion--the passivity instability criterion -- the following passivity factors, $\mu_m(\mathcal{G}_1)$ and $\mu(\mathcal{N})$ appear:

$$\int_0^T (\mathcal{G}_1 f)(t)f(t)dt \geq \mu_m(\mathcal{G}_1)\ \|f_T\|^2 \quad \text{for all } f \in M,$$

$$\int_0^T (\mathcal{N} f)(t)f(t)dt \quad \mu(\mathcal{N})\ \|f_T\|^2 \quad \text{for all } f \in L_2.$$

Note once again the different spaces in which the passivity factors are defined.

(c) Comments. (i) The instability criteria (assuming that they are valid) cannot be used for a stable $\mathcal{G}_1$. Surely, even the linear time-invariant feedback system (2.17) could become unstable for some K in $[0,\infty)$ for a stable $\mathcal{G}_1$.

(ii) Proofs of the instability criteria are by contradiction. Let us analyze the proof of the small gain instability theorem [T 1(a), p 633] by referring to Fig.5.1. Here the feedback gain (denoted by $\mathcal{N}$) is time invariant.

We assume that $\sigma \in L_2$. This implies that $\mathcal{N}\sigma \in L_2$ and hence $v(\cdot) = f(\cdot) - \mathcal{N}\sigma(\cdot)$ also belongs to L_2. But by virtue of the assumption that inputs to $\mathcal{G}_1$ in M, the proper subspace of L_2, alone give rise to outputs in L_2, we can conclude that $v \in M$. That is all. However, Takeda and Bergen proceed further and assert that we can pick $f \in M^{\perp}$ so that f and v are orthogonal. It should be observed that the space M has been defined with respect to $\mathcal{G}_1$ as an isolated block, not with respect to $\mathcal{G}_1$ in a feedback loop. That is to say, $f \in M^{\perp}$ cannot be straightaway chosen so as to give rise to the desired σ disregarding the set of system equations

$$v(\cdot) = f(\cdot) - \mathcal{N}\sigma(\cdot)$$

$$\sigma(\cdot) = \mathcal{G}_1 v(\cdot)$$

Hence with $\sigma \in L_2$, we can at the most infer that $f(\cdot) \in L_2$. Further $f(\cdot) \in L_2$ should be such that the relation

$$f(\cdot) = v(\cdot) + \quad \sigma(\cdot)$$

holds. It is erroneous to start with the assumption that $f(\cdot)$ can be chosen to lie in $M^{\perp}$.

(iii) Intuition suggests that the gains, defined below, of operators over different spaces, are not the same. (See Definition 5.42, Sec. 5.4, pp 150-151).

$$\gamma(\mathcal{G}) = \sup_{\substack{x \in L_2 \\ x \neq 0}} (\|\mathcal{G}x\| / \|x\|) \quad ;$$

$$\gamma_m(\mathcal{G}) = \sup_{\substack{x \in M \\ x \neq 0}} (\|\mathcal{G}x\| / \|x\|).$$

In fact, $\gamma_m(G) < \gamma(G)$ if M is a proper subspace of L_2. Further, a similar argument holds for the passivity factor defined over different spaces :

$$\mu(\mathcal{G}) = \inf_{\substack{x \in L_2 \\ x \neq 0}} \left(\int_0^{\infty} (\mathcal{G}x)(t)x(t)dt / \|x\|^2 \right) \quad ;$$

$$\mu_m(\mathcal{G}) = \inf_{\substack{x \in M \\ x \neq 0}} \left(\int_0^{\infty} (Gx)(t)x(t)dt / \|x\|^2 \right)$$

from which $\mu_m(G) > \mu(G)$ if M is a proper subspace of L_2. Consequently, the assertion of Takeda and Bergen [T 1(a), p 634] that the gain and the passivity factor of an unstable $\mathcal{G}_1$ over a subspace of L_2 can be found by the same techniques as in the case of a strictly stable $\mathcal{G}_1$ is untenable.

Remark 7.31. The instability criterion of Takeda and Bergen [T 1(b)] is, according to these authors, an improvement over the criterion of Willems [W 3(b)] in that there is no requirement of invertibility of the nonlinear feedback operator. However, the comments on Willems' work made in item (1) above apply here as well.

(4) Steding and Bergen [S 10] .

(a) Assumptions. $\mathcal{G}_1$ is the composition (or sum) of an infinite dimensional, stable component and a finite dimensional unstable component. Poles of the unstable component on the $j\omega$-axis are excluded.

(b) Main idea. The definition of the space M is the same as in item (3) above. The Lyapunov-Chetaev instability lemma is used in an extended form because of the infinite dimensional component of $\mathcal{G}_1$.

(c) Comments. See Comments (i) and (iii) in item (3) above.

7.4. Conclusions.

We have attempted an analysis of the L_2-instability of a feedback system described by the convolution integral equation (2.7) and also of the operator feedback system described by (7.40). The instability criteria, based on Assumption U2 (Sec.2.32, p 32) for the system (2.7) and derived using the converse Cauchy-Schwarz inequality, are counterparts of the stability criteria of Chapter 5. The criteria concerning the system (2.7) are believed to be superior to those in the literature at least in one respect : The Nyquist instability criterion for the linear time invariant feedback system (2.17) can be obtained as a special case. As regards the operator feedback system (7.40), the L_2-instability criterion is the counterpart of the wellknown passivity L_2-stability criterion of Zames [Z 3(b)] .

The results of the chapter along with those of Chapter 6 are better treated as tentative in need of further investigation. This is partly due to the fact that it is not known how to derive L_2-instability criteria for

(i) the system (2.7) under Assumption S2 (Sec.2.32, p 31) ;

(ii) the system (7.40) under a corresponding stability assumption,

and partly to the conflicting nature of the instability results found in the literature. Of course, there is the perennial problem of reproducing, on the basis of the results of the present and the last chapters, the instability boundaries of the Mathieu and Hill equations.

Appendix 7.1

Proof of Lemma 7.22. We have

$$I_{n1} = \int_0^T h(t)k(t)\exp(-2\beta t)\varphi(\sigma(t)) \Big\{(a_o - b_o\beta)\sigma(t) + b_o(d\sigma/dt)\Big\}\, dt + b_o h(0)\, \Phi(\sigma(0)) \tag{7.56}$$

By virtue of hypothesis (i), an application of the second mean value theorem (Lemma 4.31) yields the result that there exists a T' in $[0,T]$ such that

$$I_{n1} = h(0)k(0)\Big\{(a_o - b_o\beta)\int_0^{T'} \varphi(\sigma(t))\sigma(t)dt + b_o\int_0^{T'} \varphi(\sigma(t))(d\sigma/dt)dt\Big\} + b_o h(0)k(0)\,\Phi(\sigma(0)). \tag{7.57}$$

Integrating the second integral in (7.57) by parts, we get

$$I_{n1} = h(0)k(0)\Big\{(a_o - b_o\beta)\int_0^{T'} \varphi(\sigma(t))\sigma(t)dt + b_o\,\Phi(\sigma(T'))\Big\}$$

from which we conclude that I_{n1} is nonnegative for $a_o \geq b_o\beta$. The lemma is proved.

Appendix 7.2

Proof of Lemma 7.23. We have

$$I_{n2} = \int_0^T h(t)k(t)\exp(-2\beta t)\varphi(\sigma(t))\Big\{(a_o - \beta b_o)(\mathcal{G}f)(t) + b_o(d(\mathcal{G}f)(t)/dt)\Big\}\, dt$$

which, by virtue of hypothesis (i), becomes

$$I_{n2} = h(0)k(0)\int_0^{T'} \varphi(\sigma(t))\left\{(a_o-\beta b_o)(\mathcal{G}f)(t)+ b_o(d(\mathcal{G}f)(t)/dt)\right\}dt$$

where T' is some number in $[0,T]$. Note that no assumption as to the boundedness of $\mathcal{G}$ has been made. Since, by virtue of hypothesis (ii), I_{n2} is positive for all $\sigma_T(\cdot)\in L_2(0,\infty)$ and $f_T(\cdot)\in L_2(0,\infty)$, for instance, satisfying the system equation (2.7), we invoke the converse Cauchy-Schwarz inequality to conclude that

$$I_{n2} \geq \eta \, \|\sigma_{T'}\| \cdot \|f_{T'}\|$$

for some $\eta > 0$ and $f_T(\cdot)$ in a nontrivial subspace of $L_2(0,\infty)$. The lemma is proved.

Appendix 7.3

Proof of Lemma 7.24. We have

$$I_{n3} = \int_0^T h(t)k(t)\exp(-2\mu t)\exp((2\mu-\beta)t)(\sigma_\beta(t) + (\mathcal{Z}_{ac}\sigma_\beta)(t))dt \tag{7.58}$$

By virtue of hypothesis (iii), an application of the second mean value theorem (Lemma 4.31) yields the result that there exists a T' in $[0,T]$ such that

$$\begin{aligned} I_{n3} &= h(0)k(0)\int_0^{T'} \exp((2\mu-\beta)t)\varphi(\sigma(t))(\sigma_\beta(t) + (\mathcal{Z}_{ac}\sigma_\beta)(t))dt \\ &= h(0)k(0)\int_0^{T'} \exp(2(\mu-\beta)t)\varphi(\sigma(t))\sigma(t)dt + \end{aligned}$$

$$h(0)k(0)\int_0^{T'} \exp((2\mu-\beta)t)\varphi(\sigma(t))\left(\int_{-\infty}^0 z_{ac}(\tau)\sigma(t-\tau)\exp(-\beta(t-\tau))d\tau\right)dt \tag{7.59}$$

Assuming an interchange of the integration operations to be permissible in the second term on the right hand side of (7.59), we get, after some obvious manipulations,

$$I_{n3} = h(0)k(0)\int_0^{T'} \exp(2(\mu-\beta)t)\varphi(\sigma(t))\sigma(t)dt +$$

$$h(0)k(0)\int_{-\infty}^0 z_{ac}(\tau)\exp((2\mu-\beta)\tau)\left\{\int_0^{T'} \exp(-2(\beta-\mu)(t-\tau))\varphi(\sigma(t))\sigma(t-\tau)dt\right\}d\tau \tag{7.60}$$

Invoking an intermediate result in the proof of Lemma 4.36 (Appendix 4.2, p 129), we find that, for $\mu < \beta$,

$$\int_0^{T'} \exp(-2(\beta-\mu)(t-\tau))\varphi(\sigma(t))\sigma(t-\tau)dt \leq (1+\delta_s-\delta_m) \int_0^{T'} \exp(-2(\beta-\mu)t)$$

$$\varphi(\sigma(t))\sigma(t)dt$$

for all $\sigma(\cdot)$. Hence, by virtue of hypotheses (i) and (ii), we conclude from (7.60) that I_{n3} is nonnegative for all $\sigma(\cdot)$ in the domain of $\mathcal{Z}_{ac}$, for all $T > 0$. The lemma is proved.

Appendix 7.4

Proof of Lemma 7.25. We have, by virtue of hypothesis (i),

$$I_{n4} = h(0)k(0) \int_0^{T'} \exp(2(\mu-\beta)t)\varphi(\sigma(t)) \left\{ (\mathcal{G}f)(t) + (\mathcal{Z}_{ac}^{\beta} \mathcal{G}f)(t) \right\} dt \qquad (7.61)$$

where T' is some number in $[0,T]$. Further, since $\mu < \beta$, we conclude from (7.61) that

$$I_{n4} = h(0)k(0) \int_0^{T''} \varphi(\sigma(t)) \left\{ (\mathcal{G}f)(t) + (\mathcal{Z}_{ac}^{\beta} Gf)(t) \right\} dt$$

where T'' is some number in $[0,T']$. Note that no assumption of boundedness on $\mathcal{G}$ has been made.

By virtue of hypothesis (ii), I_{n4} is positive for all $\sigma_T(\cdot) \in L_2(0,\infty)$ and $f_T(\cdot) \in L_2(0,\infty)$, for instance, satisfying the system equation (2.7). We invoke the converse Cauchy-Schwarz inequality to conclude that

$$I_{n4} \geq \eta \, \|\sigma_T\| \cdot \|f_T\|$$

for some $\eta > 0$ and $f_{T''}(\cdot)$ in a nontrivial subspace of $L_2(0,\infty)$. The lemma is proved.

EXERCISES

7.1. Indicate how one can derive the instability counterparts of the results of Chapter 4.

7.2. What is the multiplier function form of the L_2-stability condition for the linear time invariant system (2.17) ? Hence derive the multiplier function form of the Basic Result 7.21. Analyze its role in generalizing, if possible, the instability results of Sec.7.22 and Sec.7.23.

7.3. Derive an L_2-instability criterion (analogous to Theorem 7.21) for the linear time varying system (2.23) using $z_c \in \mathcal{L}$.

7.4. Interpret Theorem 7.21 and the criterion obtained in Exercise 7.3 geometrically. Compare with the corresponding instability criterion (Criterion 6.51) of Chapter 6.

7.5. Complete the proofs of Theorems 7.22 and 7.23, and comment on the existence of a counterpart to the result mentioned in Exercise 5.2 (p 158). Also derive, based on Theorem 7.24 or on the finite gain versions (like circle criterion), counterparts to the results mentioned in Exercise 5.3 (p 158). Repeat the Exercise for Lemma 5.42 (p 153).

7.6. Using the multiplier operators $1 + z_c$, $1 + z_c + z_{ac}$ belonging to class $\mathcal{L}$, derive instability criteria for (2.7) with $\varphi(\cdot) \in C_m$. Interpret the criteria geometrically. How can one derive L_p-instability criteria , which are counterparts to those asked for in Exercise 5.11 (p 160) ?

7.7. How can one establish the necessity or otherwise of the L_2-instability conditions for the operator feedback system (7.40) as given in Theorem 7.24 ?

7.8. Analyze the discrepancy (see Remark 7.24) between the time-domain bound on the integral involving $z_{ac}(t)$ in Theorem 7.23 and that on the integral involving $z_g(t)$ in Theorem 6.54. Suggest means of improving the latter in consonance with the former.

7.9. Consider the problem of analyzing the L_2-instability of the system of Exercise 5.12 (p 160) described by a linear partial differential equation. Suggest suitable classes of input and output functions for which an instability definition can be given. Repeat the exercise for the system of Exercise 5.14 (p 161) described by a nonlinear partial differential equation.

EPILOGUE

Stability and instability analyses constitute the main requirements in the synthesis of complex systems, of which the basic nonlinear time varying feedback system considered in the monograph forms the prototype. When it is noted that all the classical design techniques for linear time invariant models, which are obtained as a gross simplification of the original systems, are inspired by stability and instability considerations, new synthesis techniques, based on the stability and instability results of the type found in the monograph, should lead to a more realistic design of nonlinear time varying systems.

The mathematical description of the basic nonlinear time varying feedback system is in one of the three forms specified below, and, correspondingly, the methods employed for the stability and instability analyses are different :

1. Differential equation (with no input) ;
2. Integral equation with input ; and
3. More general integral equation with input.

The methods employed in stability and instability analyses, and the nature of the solution behaviour of the system so established are respectively as follows :

1. Lyapunov - Corduneanu. Exponential stability and instability;
2. Popov. Exponential stability and instability; and
3. Zames - Sandberg. L_2-stability and instability.

An overview of the methods brings out the interesting common feature that a certain quantity related to the 'generalized energy' of the system is evaluated along the trajectories of the system. The success of each of the methods rests on the fact that the generalized energy , suitably chosen and evaluated along the trajectories of the system, is governed by an inequality (differential, integral or norm in character corresponding to the method employed) which is simpler to resolve than the original set of equations. The behaviour of the generalized energy established in this fashion reflects the behaviour of the solutions of the systems under consideration. Hence the title of the book.

As is the situation in any mathematical study of systems, certain assumptions about the known behaviour of a special case (of the original system) form the starting point for stability and instability analyses. From an application of all the three methods, it turns out that

(i) a very useful special case is the linear time invariant feedback system ;

(ii) the assumptions made for stability analysis of the nonlinear time varying feedback system are counterparts of those for instability analysis of the same system ;

(iii) identical assumptions for both stability and instability analyses lead to no definitive conclusions on the necessity or otherwise of of the stability and instability conditions ; and, more importantly,

(iv) the stability and instability criteria cannot reproduce, in the case, of the (linear second order) Mathieu and Hill equations, the well known stability and instability boundaries..

The only superiority claimed for results of the type found in the monograph is that linear and nonlinear time varying systems of an order higher than two (which are amenable to representation by the structure of Fig 2.1) can be analysed for stability and instability. However, it is a relief to find that, at a time when a research worker is on a treadmill of abstraction involving causality and invertibility and forced to run as fast as he can to stay even,' a linear second order time varying system cannot be handled convincingly and conclusively by the general methods of stability and instability analyses. Here is the clarion call to the next Lyapunov, Popov, Zames or Sandberg or all combined in one.

ERRATA

Notation : In the second column, unless the item is explicitly an equation or a Table, the first component is the number of the line, the second component stands for 'from the top' (denoted by 'a'), 'from the bottom' (denoted by 'b'), 'below the equation/material' (denoted by 'c' and the equation number/material), or 'above the equation' (denoted by 'd' and the equation number).

Page No.	Location	Material	To be read as (unless otherwise indicated)
5	3 to 6 c (1.9)	Conversely,···, a long time ago.	Delete (in view of the results of Chapter 6).
7	4 c Vector equation	equation	equations
10	2 d (1.11)	prevoiusly	previously
13	3 a	on character of	on the character of
21	9 b	coefficient	coefficients
23	6 b	everywhere.)	everywhere).
25	(2.5b)	$\exp(A_o(t-t_o)\underline{x}_o$	$\exp(A_o(t-t_o))\underline{x}_o$
26	6 a 11 a 12 a	(2,1) Fig.1 Fig.3	(2.1) Fig.2.1 Fig.2.3
30	7 a	equaivalently	equivalently
36	7 b	Fig.2.1	Fig.2.2
37	10 a	That is	That is,
40	6 a	related	treated

Page No.	Location	Material	To be read as (unless otherwise indicated)
43	11 b	is used	along the solutions of (2.1) or (2.4), is used.
45	7 b 5 b	candidates Problem)2.42)	conditions Problem 2.42)
47	5 b	... is that	nificant that
48	11 a	saability	stability
54	9 b	causal and non-causal	causal and anticausal
65	3 b	constane	constant
74	8 b	$V_1($ $,t)$	$V_1(\underline{x},t)$
79	10 a	Re a_o +	Re (a_o +
90	11 a 14 a	analyses $\alpha s^2+1)^2$	analysed $(\alpha s^2+1)^2$
100	2 c Table	Table	Table 4.2
105	1 d (4.28) 1 c (4.29)	Z^a_{ni} γ_{ai}	Z^a_{nl} γ_{al}
110	1 b	funcdamental	fundamental
116	1 c item 9	Will ms	Willems
121	12 a	Propoerty	Property
123	2 a 15 b	(except an	except and
134	2 a	a $t_o \geq 0$	all $t_o \geq 0$

Page No.	Location	Material	To be read as (unless otherwise indicated)
137	3 b	stablish	Establish
140	4 a 7 b	obsevable interest	observable of interest
144	1 c (5.2)	and $\{z_i\}$	and $\{z_i'\}$
145	(5.4)	$\exp(\gamma_{ac}\tau)$	$\exp(-\gamma_{ac}\tau)$
150	2 a	by virtue o	by virtue of
153	15 a	(5.18) become	(5.18) becomes
156	10 a	mark 5.42	Remark 5.42
157	14 b	sicists Continue	Physicists Continue
170	2 d (6.18)	satifies	satisfies
175	(6.28)		add the term $b_o\Phi(\sigma(0))$
180	7 b	oovious	obvious
181	2 a	. $\varphi(\cdot)$	$\varphi(\cdot)$
185	3 b	Lemma 6.31	Lemma
186	9 a	$G(j\omega\ \beta)$	$G(j\omega+\beta)$
192	(6.79)	$\exp(-\tau\alpha/2)$	$\exp(-\tau\alpha/2)\Phi($
226	8 b	gcneralized	'generalized
227	5 b	and forced	and 'forced

ADDITIONAL REFERENCES

W 3. Willems, J.C., (e) Some results on the L_p-stability of linear time varying systems, IEEE Trans.Automatic Control, AC-14(1969), pp 660-665.

F 6. Freedman, M.I., Falb, P.L., and Zames, G., A Hilbert space stability theory over locally compact abelian groups, SIAM J.Control 7(1969), pp 479-495.

C 9. Cook, P.C., Circle criteria for stability in Hilbert space, SIAM J. Control, 13(1975), pp 593-610.

F 7. Fitts, R.E., Two counter-examples to Aizerman's conjecture, IEEE Trans. Automatic Control, AC-11(1966), pp 553-556.

Y 4. Yakubovich, V.A., and Starzhinskii, V.M., Linear Differential Equations with Periodic Coefficients, Vols.1 and 2, John Wiley and Sons (Israel Program for Scientific Translations), 1975.

REFERENCES

A 1. Acker, A., Stability criteria for time varying systems in Hilbert space, SIAM J.Control, 13(1975), pp 1156-1171.

A 2. Aizerman, M.A., On a problem concerning the stability in the large of dynamical systems, Uspekhi Mat. Nauk, 4(1949), pp 187-188.

A 3. Aizerman, M.A., and Gantmacher, F.R., Absolute Stability of Regulator Systems, Holden-Day, San Francisco, 1964.

A 4. Anderson, B.D.O., External and internal stability of linear systems — a new connection, IEEE Trans. Automatic Control, AC-17(1972), pp 107-111.

A 5. Antosiewicz, H.A., (a) A note on asymptotic stability, Quart. Appl. Math. 9(1951), pp 317-319,(b)Int.Conf.Diff.Equations, Acad.Press 1975.

B 1. Baker, R.A., and Bergen, A.R., On the relation between absolute stability with respect to a set of linear time varying elements and a set of nonlinear time varying elements, IEEE Trans. Automatic Control AC-15(1970), pp 503-504.

B 2. Bell, C.V., and Wade,G., Iterative travelling wave parametric amplifiers, IEEE Trans. Circuit Theory, CT-7(1960), pp 4-11.

B 3. Barbashin, E.A., and Krasovskii, N.N., On the stability of motion in the large, Dokl. Ak. Nauk SSSR, 86(1952), pp 453-454.

B 4. Bellman, R., (a) Stability Theory of Differential Equations, McGraw-Hill, 1953 ; (b) Converses of Schwarz inequality, Duke Math. J. 23 (1956), pp 429-434 ; (c) An Introduction to Matrix Analysis, McGraw-Hill, 1960.

B 5. Bellman, R., and Beckenbach, E.F., Inequalities, Springer, 1965.

B 6. Bergen, A.R., Iwens, R.P., and Rault, A.J., On the stability of nonlinear feedback systems, IEEE Trans.Aut.Cont. AC-11(1966),pp 742-744.

B 7. Bergen, A.R., and Rault, A.J., Absolute input-output stability of feedback systems with a single time varying gain, Journal of the

Franklin Institute, 286(1968), pp 280-294.

B 8. Bermant, M.A., and Yemelyanov, S.V., Stability of a class of automatic control systems with variable parameters, Engineering Cybernetics, No.6, 1964, pp 147-153

B 9. Bhatia, N.P., On exponential stability of linear differential systems, J.SIAM Control Ser A. 2(1965), pp 181-191.

B 10. Bickart, T.A., Periodically time varying system, an instability criterion, Proc.IEEE 55(1967), pp 2057-2058.

B 11. Birkhoff, G., (a) Stability of spherical bubbles, Quart.Appl.Math. 13(1956), pp 451-453; (b) Note on Taylor instability, Quart.Appl. Math. 12(1954), pp 306-309.

B 12. Blodgett, R.E., and Young, K.P., (a) On the stability of a second order time varying nonlinear system, IEEE Trans. on Automatic Control, AC-16(1971), pp 270-272 ; (b) A combined time frequency condition for stability of time varying systems with one nonlinearity, Trans.ASME (J.Dynamic Syst.Meas.Contr.), 93(1971),pp 261-268.

B 13. Bolotin, V.V., The Dynamic Stability of Elastic Systems, Holden-Day, California 1964.

B 14. Bongiorno, J.J.Jr., (a) Stability and convergence properties of model reference adaptive control systems, IRE Trans.on Automatic Control, AC-7(1962), pp 30-41 ; (b) Real frequency stability criteria for linear time varying systems, Proc.IEEE, 52(1964), pp 832-841.

B 15. Borg, G., Über die Stabilität gewisser Klassen von linearen Differential gleichungen, Arkiv för Matematik, Astronomi och Fysik, Band 31A(1944), pp 1-31.

B 16. Brockett, R.W., (a) The status of stability theory for deterministic systems, IEEE Trans. on Automatic Control, AC-11(1966) pp 596-

606; (b) On the stabilty of nonlinear feedback systems, IEEE Trans. on Applications and Industry, 75(1964), pp 443-449 ; (c) Optimization theory and the converse of the circle criterion, Proc.NEC 21 (1965), pp 697-701 ; (d) Variational methods for the stability of periodic equations in Differential Equations and Dynamical Systems edited by J.Hale and J.P.LaSalle, Academic Press, New York, 1967, pp 299-308 ; (e) Finite Dimensional Linear Systems, John Wiley, New York, 1970.

B 17. Brockett, R.W., and Forys, L.J., On the stability of systems containing a time varying gain, Proc.Second Allerton Conference on Circuit and System Theory, 1964, pp 413-430.

B 18. Brockett, R.W., and Lee, H.B., Frequency domain instability criteria for time varying and nonlinear systems, Proc.IEEE, 55(1965), pp 604-619.

B 19. Brockett, R.W., and Skoog, R.A., Impedance synthesis using time varying resistors and capacitors, Proc. Second Annual Princeton Conf. on Information Sciences and Systems, 1968.

B 20. Brockett, R.W., and Willems, J.L., Frequency domain stability criteria, Part I, IEEE Trans. on Automatic Control, AC-10(1965), pp 255-261 ; Part II, AC-10(1965), pp 407-413.

B 21. Burton, T.A., On the construction of Liapunov functions, SIAM J. Appl.Math. 17(1969), pp 1078-1085.

C 1. Carroll, R.L., Application of Liapunov-model tracking parameter identification for the CH-47 helicopter, Proc.IEEE 1975, Decision and Control Conference, pp 858-863.

C 2. Cesari, L., Asymptotic Behaviour and Stability Problems in Ordinary Differential Equations, Second Edition, Academic Press, New York,1963.

C 3. Chen, C.T., (a) Bounded-input bounded-output stability of linear time varying feedback systems, J.Franklin Institute, 286(1968), pp 123-126 ; (b) L_p- stability of linear time varying feedback systems, SIAM J.Control, 6(1968), pp 186-192.

C 4. Cho, Y.S., and Narendra, K.S., An off-axis circle criterion for the stability of feedback systems with a monotonic nonlinearity, IEEE Trans.on Automatic Control, AC-13(1968), pp 413-416.

C 5. Chow, S.N., and Dunninger, D.R., Lyapunov functions satisfying $\dot{V} > 0$, SIAM J.Appl.Math. 26(1974), pp 165-168.

C 6. Coddington, E.A., and Levinson, N., Theory of Ordinary Differential Equations, McGraw Hill, New York, 1955.

C 7. Cooley, W.W., Clark, R.N., and Buckner, R.C., Stability in linear systems having a time variable parameter, IEEE Trans. on Automatic Control, AC-9(1964), pp 426-434.

C 8. Corduneanu, C., Applications of differential inequalities to the theory of stability, Anal.Sti.Univ.Al.I.Cuza, Iasi(Serie noua), Sec.I, 6(1960), pp 47-58. See also Mathematical Reviews 23(1962), A2604.

D 1. Damborg, M.J., Stability of the Basic Nonlinear Operator Feedback System, Ph.D. Thesis, The University of Michigan, USA 1969.

D 2. Damborg, M.J., and Naylor, A.W., Stability structure for feedback systems having unstable loops, IEEE Trans.on Automatic Control, AC-18 (1973), pp 318-319.

D 3. Davis, J.H., (a) Mean square gain criteria for the stability and instability of time varying systems, IEEE Trans.on Automatic Control, AC-17(1972), pp 214-219 ; (b) Encirclement conditions for stability and instability of feedback systems with delays, Int.J.Control, 15 (1972), pp 793-799.

D 4. Desoer, C.A., (a) A general formulation of the Nyquist criterion, IEEE Trans.on Circuit Theory, CT-12(1965), pp 230-234 ; (b) A generalization of the Popov criterion, IEEE Trans.on Atomatic Control, AC-10(1965), pp 182-185.

D 5. Desoer, C.A., Liu, R., and Auth, L.V., Lineraity vs nonlinearity and asymptotic stability in the large, IEEE Trans.on Circuit Theory, CT-12(1965), pp 117-118.

D 6. Desoer, C.A., and Vidyasagar, M., Feedback Systems : Input-Output Properties, Academic Press, New York 1975.

D 7. Donalson, D.D., and Leondes, C.T., Parameter tracking for adaptive control, IEEE Trans.on Applications and Industry, 68(1963), p 241.

E 1. Elmaraghy, R., and Tabarrok, B., On the dynamic stability of an axially oscillating beam, J.Franklin Institute, 300(1975), pp 25-39.

E 2. El'sin, M.I., A solution to a classical problem of oscillations, Soviet Math.Dokl. 3(1962), pp 1013-1016.

E 3. Erugin, N.P., Theorems on instability, Prikl.Math.i Mekh., 16(1952), pp 355-361.

F 1. Falb, P.L., and Zames, G., On cross correlation bounds and positivity of certain nonlinear operators, IEEE Trans.on Automatic Control, AC-12(1967), pp 219-221.

F 2. Farrell, J.L., and Gundersdorf, C., Time varying Doppler navigation servo, IEEE Trans.on Automatic Control, AC-9(1964), pp 282-285.

F 3. Fearnsides, J.J., and Levine, W.S., On the determination of the asymptotic behaviour of an inertially oriented space station, IEEE Trans. on Automatic Control, AC-19(1974), pp 186-191.

F 4. Franks, L.E., and Sandberg, I.W., An alternative approach to the realization of network transfer functions : The N-path filter, Bell

System Technical Journal, 39(1960), pp 1321-1350.

F 5. Freedman, M.I., (a) Phase function norm estimates for stability of systems with monotone nonlinearities, SIAM J.on Control, 10(1972), pp 99-111; (b) L_2-stability of time varying systems - construction multipliers with prescribed phase characteristics, SIAM J.on Control, 6(1968), pp 559-578.

F 6. Freedman, M.I., and Zames, G., Logarithmic variation criteria for the stability of systems with time varying gains, SIAM J.on Control, 6(1968), pp 487-507.

G 1. Gantmacher, F.R., Theory of Matrices, Vols 1 and 2, Chelsea Publishing Company, New York 1960.

G 2. Gardner, W.A., Modulation rate distortion in frequency modulators, IEEE Trans.on Circuit Theory, CT-16(1969), pp 295-302.

G 3. Genin, J., and Maybee, J.S., Boundedness theorem for a nonlinear Mathieu equation, Quart.Appl.Math.28(1970), pp 450-453.

G 4. Gohberg, I.C., and Krein, M.G., On factorization of operators in Hilbert space, Soviet Math.Dokl. 3(1962), pp 1578-1582.

G 5. Gruber, M., and Willems, J.L., A generalization of the circle criterion, Proc. Fourth Allerton Conf.on Circuit and System Theory, Univ. Illinois, Urbana, Illinois 1966, pp 827-835.

G 6. Guillemin, E.A., Synthesis of Passive Networs, New York, Wiley, 1957.

G 7. Gunderson, H., Rigas, H., and Van Vleck, F.S., A technique for determining stability regions for the damped Mathieu equation, SIAM J.Appl. Math. 26(1974), pp 345-349.

H 1. Hacker, T., Stability of partially controlled motions of an aircraft, Journal of the Aerospace Sciences , 28(1961), pp 15-26 ; (b) Flight Stability and Control, American Elsevier, 1970.

H 2. Haddad, E.K., (a) New criteria for bounded-input bounded-output and asymptotic stability of nonlinear systems, Paper 32-2, Proc.IFAC Conference in Paris 1972 ; (b) Stability of nonlinear feedback systems with time varying linear sub system (plant), Int.J.on Control, 18(1973), pp 1077-1103.

H 3. Hahn, W., (a) Theory and Application of Liapunov's Direct Method, Prentice Hall, Englewood Cliffs, N.J., 1963 ; (b) Stability of Motion, Springer Verlag, Berlin 1967..

H 4. Halanay, A., Differential Equations : Stability, Oscillations, Time Lags, Academic Press New York, 1966.

H 5. Harvey, C.A., Limit cycle bounds of a satellite attituda control system, IEEE Trans.on Automatic Control, AC-17(1972), pp 526-528.

H 6. Hayashi, C., Nonlinear Oscillations in Physical Systems, Mc Graw Hill, New York, 1964.

H 7. Hiza, J.G.,and Li, C.C., Analytical synthesis of a class of model reference time varying control systems, IEEE Trans. on Appl. and Industry, November 1963, pp 356-362.

H 8. Hobson, E.W., The Theory of Functions of a Real Variable,Cambridge University Press 1927.

H 9. Holtzman, J.M., Nonlinear System Theory, Prentice Hall, Englewood Cliffs, N.J., 1970.

H 10. Hsu, J.C.. and Meyer, A.U., Modern Control Principles and Applications Mc Graw Hill Book Company, New York, 1968.

I 1. Infante, E.F., On the stability of linear autononous random systems, Trans.ASME J.Appl.Mech. 36(1968), p 7.

I 2. Infeld, E., On the stabilty of solutions of a second order differential equation, Quart.Appl.Math. 32(1975), pp 465-467.

I 3. Infeld, E., and Rowlands, G., On the stability of nonlinear cold plasma waves II, J.Plasma Physics, 10(1973), p 233.

J 1. Johnson, C.D., A note on control systems with one nonlinear element, IEEE Trans.Automatic Control, AC-11(1966), pp 122-124.

J 2. Jones,Jr., J., On the asymptotic stability of certain second order nonlinear differential equations, J.SIAM Appl.Math. 14(1966), pp 16-22.

K 1. Kalman, R.E., (a) Physical and mathematical mechanisms of instability in nonlinear automatic control systems, Trans.ASME, 79(1957), pp 553-566; (b) Lyapunov functions for the problem of Lur'e in automatic control, Proc.Nat.Acad.Sci. USA 49(1963), pp 201-205. (c) On the stability of time varying linear systems, IRE Trans.on Circuit Theory, CT-9(1962), pp 420-422 ; (d) Mathematical description of linear dynamical systems, SIAM J.Control, 1(1963), pp 152-192.

K 2. Kalman, R.E., and Bertram, J.E., Control system design via the second method of Liapunov, Part I. Continuous systems, Trans.ASME, J.Basic Engineering, 82(1960), pp 371-393.

K 3. Kane, T.R., and Barba, P.M., Attitude stability of a spinning satellite in an elliptic orbit, Trans.ASME, J.Appl.Mech. 33(1966), pp 402-405.

K 4. Kaplan, W., Operational Methods for Linear Systems, Addison-Wesley, Publishing Company, Reading, Massachusetts, 1962.

K 5. Keenan, R.K., Parametrically induced unstable oscillations in higher order circuits, Proc.IEEE (Letters), 56(1968), p 1395.

K 6. Komkov, V., On boundedness and oscillation of the differential equation, $x'' + A(t)g(x) = f(t)$ in R^n , SIAM J.Appl.Math. 22(1972), pp 561-568.

K 7. Krasovskii, N.N., Stability of Motion, Stanford University Press, Stanford, Calif. 1963.

K 8. Kuh, E.S., Stabilty of linear time varying networks - state space approach, IEEE Trans.on Circuit Theory, CT-12(1965), pp 150-157.

L 1. Lakshmikantham, V., and Leela, S., Differential and Integral Inequalities : Theory and Applications, Vol 1, Academic Press 1969.

L 2. Lal, M., Dixit, S.B., and Avasthy, P.N., A stability criterion for periodically varying systems, Proc.IEEE (Letters), 57(1969), pp 226-227.

L 3. La Salle,J.P., Stability and control, SIAM J.Control 1(1962),pp 3-15.

L 4. LaSalle, J.P., and Lefschetz, S., Stability by Liapunov's Direct Method, Academic Press, New York, 1961.

L 5. Levy, D.N., and Keller, J.B., Instability intervals of Hill's equation Comm.Pure Appl.Math. 16(1963), pp 469-476.

L 6. Lowis, O.J., The stability of a rotor blade flapping motion at high tip speed ratios, Aeronautical Research Council, 1962, Paper 23, p 371

L 7. Lubkin, S., and Stoker, J.J., Stability of columns and strings under periodically varying forces, Quart.Appl.Math. 1(1943), pp 215-236.

L 8. Lur'e, A.I., Some Nonlinear Problems in the Theory of Automatic Control, Her Majesty's Stationary Office, London, 1957.

L 9. Lur'e, A.I., and Postnikov, V.N., On the theory of stability of control systems, Prikl.Math.i.Mekh. 8(1944), pp 246-248.

L 10. Lyapunov, A.M., Problème générale de la stabilité du mouvement, Annals of Mathematical Studies 17, 1949.

M 1. MacCamy, R.C., and Wong, J.S.W., Stability theorems for some functional equations, Trans. Amer.Math.Soc.164(1972), 1-37.

M 2. MacDonald, J.R., and Edmundson, D.E., Exact solution of a time varying capacitance problem, Proc.IRE 49(1961), pp 453-466.

M 3. Mack, J.W., An example of the theory of nonlinear oscillations, SIAM J.Appl.Math. 17(1969), pp 516-519.

M 4. Magnus, W., and Winkler, S., Hill's Equation, Wiley, New York 1966.

M 5. Markus, L., and Yamabe, H., Global stability criteria for differential systems, Osaka Math,J. 12(1960), pp 305-317.

M 6. Maxwell, J.C., On governors, Proc.Royal Soc.London 16(1868), pp 270-280.

M 7. McLachlan, N.W., Ordinary Differential Equations in Engineering and Physical Sciences, Oxford, Clarendon Press, 1956.

M 8. Mel'nikov, G.I., Some questions in Lyapunov's direct method, Dokl.Ak. Nauk SSSR, 110(1956), pp 326-329 (Russian).

M 9. Michel, A.N., Stability and trajectory behaviour of composite system, IEEE Trans.Circuits and Systems, CAS-22(1975), pp 305-312.

M 10. Minorsky, N., Self excited oscillations in dynamical systems possessing retarded action, J.Appl.Mechanics 9(1942), pp 65-71.

M 11. Mitropolskii, Yu.A., Problems of the Asymptotic Theory of Nonstationary Vibrations, Israel Program for Scientific Translations, Jerusalem 1965.

M 12. Moore, M.H., An inner product inequality, SIAM J. on Mathematical Analysis, 4(1973), pp 514-518.

M 13. Mote, C.D. Jr., Dynamic stability of an axially moving band, J.Franklin Institute, 285(1967), pp 329-346.

M 14. Murayama, T., and Ozaki, T., Feedback control of an airplane with time varying gain, Memoirs of the Defense Academy, XIII(1973), pp 357-372.

N 1. Narendra, K.S., and Goldwyn, R.M., A geometrical criterion for the stability of certain nonlinear nonautonomous systems, IEEE Trans.on Circuit Theory, CT-11(1964), pp 406-408.

N 2. Narendra, K.S., and Taylor, J.H., Frequency Domain Criteria for Absolute Stability, Academic Press, New York 1973.

N 3. Noldus, E., (a) On the stability of nonlinear systems, IEEE Trans.on Automatic Control, AC-18(1973), pp 404-405 ; (b) Instability of time varying and nonlinear feedback systems, J.Engineering Mathematics, 4(1970), pp 243-259; (c) Instability for time varying nonlinear functional differential systems, SIAM J.Control, 13(1975), pp 420-433 ; (d) Criteria for unbounded motion by positive operator methods, Int. J.Control, 18(1973), pp 289-296.

O 1. Orsic, M., and Krajcinovic, D., Resonant circuit with impulsively varying capacitance, IEEE Trans.on Circuit Theory, CT-19(1972), pp 533-536.

O 2. O'Shea, R.P., (a) An improved frequency time domain stability criterion for autonomous continuous systems, IEEE Trans.on Automatic Control, AC-12(1967), pp 725-731 ;(b) A combined frequency - time domain stability criterion for autonomous continuous systems, ibid., AC-11 (1966), pp 477-484.

O 3. Ostrowski, A., Uber Normen von Matrizen, Math.Zeitshriften, 63(1955), pp 2-18.

P 1. Parks, P.C., The circle criterion and the damped Mathieu equation, Electronics Letters, 2(1965), p 315.

P 2. Pao, C.V., On stability of nonlinear differential systems, Int.J.Nonlinear Mechanics, 8(1973), pp 219-238.

P 3. Pfaffelhuber, E.P., Generalized impulse response and causality, IEEE Trans.on Circuit Theory, CT-18(1971), pp 218-223.

P 4. Pipes, L.A., Four methods for the analysis of time variable networks, IRE Trans.on Circuit Theory, CT-2(1955), pp 4-12.

P 5. Pirogov, I.Z., On the stability of one gyroscopic system, Prikl.Mat. Mekh. 23(1959), pp 1134-1136.

P 6. Pliss, V.A., Certain Problems in the Theory of Stability of Motion in the Large, Izd.Leningradskovo University 1958.

P 7. Pontryagin, L.S., Ordinary Differential Equations, Addison-Wesley, Pergamon Press, 1962.

P 8. Popov, V.M., (a) Absolute stability of nonlinear control systems of automatic control, Automation and Remote Control, 22(1962), pp 857-875 ; (b) Dichotomy and stability by frequency domain methods, Proc. IEEE, 62(1974), pp 548-562.

P 9. Pun, L., Initial conditioned solutions of a second order nonlinear conservative differential equation with a periodically varying coefficient, J.Franklin Institute, 295(1973), pp 193-216.

P 10. Pyatnitskii, E.S., (a) Absolute stability of nonstationary nonlinear systems, Automation and Remote Control 31(1970), pp 1-9 ;
(b) On the existence of absolutely stable systems for which Popov's criterion is not satisfied, ibid., 32(1971), pp 22-29;
(c) New research on the absolute stability of automatic control systems,ibid., 28(1968), pp 855-881 ; (d) Criterion for the absolute stability of second order nonlinear controlled systems with one nonlinear nonstationary element, ibid., 32(1971), pp 5-16.

R 1. Raman, C.V., Experimental investigations on the maintenance of vibrations, Proc.Indian Assoc.for the cultivation of Sci. Bulletin 6, 1912.

R 2. Ramar, K., and Ramaswamy, B., Transformation of time variable multi-input systems to a canonical form, IEEE Trans.on Automatic Control,

AC-16(1971), pp 371-374.

R 3. Reis, G.C., Stability of distributed systems with feedback via Michailov's criterion, Bell System Technical Journal, 51(1972), pp 903-919.

R 4. Rekasius, Z.V., and Rowland, J.R., A stability criterion for feedback systems containing a single time varying nonlinear element, IEEE Trans Trans.on Automatic Control, AC-10(1965), pp 352-353.

R 5. Rohrer, R.A., Stability of linear time varying networks- bounds on stored energy, Proc.NEC 19(1963), pp 107-114.

R 6. Rosenbrock, H.H., (a) Lyapunov function for some naturally occuring linear homogeneous time dependent equations, Automatica, 1(1963), pp 97-109 ; (b) The stabilty of time dependent control systems, Journal of Electronics and Control, 15(1963), pp 73-80.

S 1. Saeks, R., On the encirclement criterion and its generalization, IEEE Trans.Circuits and Systems, CAS-22(1975), pp 780-785.

S 2. Sandberg, I.W., (a) A frequency domain condition for the stability of feedback systems,containing a single time varying nonlinear element, Bell Systems Technical Journal, 43(1964), pp 1601-1608 ; (b) On the stability of linear systems of a time varying element with restricted rate of variation, IEEE Convention Record, 14(1966), Part 7, pp 173-182 ; (c) On the stability of solutions of linear differential equations with periodic coefficients, SIAM J.Appl.Math. 12(1964), pp 487-496 ; (d) A stability criterion for linear networks containing time varying capacitors, IEEE Trans.Circuit Theory, CT-12(1965), pp 2-11 ; (e) A condition for the L_∞-stability of feedback systems containing a single time varying nonlinear element, Bell System Technical Journal, 43(1964), pp 1815-1817.

S 3. Sansone, G., and Conti, R., Nonlinear Differential Equations, Mac -

millan, New York, 1964.

S 4. Silverman, L.M., and Anderson, B.D.O., Controllability, observability and stability of linear systems, SIAM J.Control, 6(1968), pp 121-130.

S 5. Skoog, R.A., Positivity conditions and instability criteria for feedback systems, SIAM J.Control, 12(1974), pp 83-98.

S 6. Skoog, R.A., and Lau, C.C.G., Instability of slowly varying systems, IEEE Trans.on Automatic Control, AC-17(1972), pp 86-92.

S 7. Skoog, R.A., and Willems, J.C., Orthogonality, positive operators and the frequency power formulas, Quart.Appl.Math. 29(1971), pp 341-361.

S 8. Srinath, M.D., Thathachar, M.A.L., and Ramapriyan, H.K., Stability of a class of nonlinear time varying systems, Int.J.Control, 7(1968), pp 117-132.

S 9. Starzhinskii, V.M., A survey of works on the conditions of stability of the trivial solution of a system of linear differential equations with periodic coefficients, Amer.Math.Soc.Translations Ser.2,1(1955), pp 189-231.

S 10. Steding, T.L., and Bergen, A.R., Instability of feedback systems using Lyapunov functionals, SIAM J.Control, 12(1974), pp 664-678.

S 11. Stevens, K.K., On linear differential equations with periodic coefficients, SIAM J.Appl.Math. 14(1966), pp 782-795.

S 12. Stoker, J.J., Nonlinear Vibrations, Interscience , New York, 1950.

S 13. Strutt, J.W., Scientific Papers, Vol 3, Cambridge University Press, Cambridge, England 1894.

S 14. Sundaresan, M.K., and Thathachar, M.A.L., (a) Time varying system stability - interchangeability of the bounds on the logarthmic variation of the gain, IEEE Trans.on Automatic Control, AC-18(1973), pp 405-407 ; (b) Average variation L_2-stability criteria for time varying feedback systems - a unified approach, ibid., AC-19(1974), pp 427-429 ; (c) L_2-stability of linear time varying systems : Conditions involving noncausal multipliers, ibid. AC-17(1972), pp 504-510; (d) L_2-stability of nonstationary feedback systems : Frequency domain criteria, ibid., AC-19(1974), pp 217-224.

T 1. Takeda, S., and Bergen, A.R., (a) Instability of feedback systems by orthogonal decomposition of L_2, IEEE Trans.on Automatic Control, AC-18(1973) , pp 631-636; (b) On the instability of feedback systems with a single nonlinear time varying gain, ibid., AC-16(1971), pp 462-464.

T 2. Thathachar, M.A.L., A stability criterion for power law nonlinearities, Automatica, 6(1970), pp 721-730.

T 3. Thathachar, M.A.L., Srinath, M.D., and Ramapriyan, H.K., On a modified Lur'e problem, IEEE Trans.on Automatic Control, AC-12(1967), pp 731-739.

T 4. Timoshenko, S., Vibration Problems in Engineering, Van Nostrand, Princeton 1928..

T 5.. Timoshenko, S.P., and Gere, J.W., Theory of Elastic Stability, McGraw-Hill Book Company, New York, 1961.

T 6. Titchmarsh, E.C., Theory of Fourier Integrals, 2 nd.ed., Clarendon Press,Oxford 1962.

V 1. Varaiya, P.P., and Liu, R., Bounded-input bounded output stability of nonlinear time varying differential systems, SIAM J.Control, 4(1966), pp 698-704.

V 2. Venkatesh, Y.V., (a) Global variation criteria for stability of linear time varying systems, SIAM J.Control, 9(1971), pp 431-440 ; (b) an exponential stability criterion for certain nonlinear time varying systems, J.Engng.Math. 8(1974), pp 125-131 ; (c) Geometric stability criteria for certain nonlinear time varying systems, Int.J.Nonlinear Mech. 10(1975), pp 245-252 ; (d) Noncausal multipliers for nonlinear system stability, IEEE Trans. Automatic Control, 15(1970), pp 195-204: (e) On the positivity of nonlinear time varying operators, ibid., AC-18(1973), pp 321-322 ; (f) On the stability of nonlinear systems with periodic coefficients, Proc.IEEE 62(1974), pp 278-279 ; (g) Alternative stability bounds on the rate of variation of the gain of feedback systems, Ricerche di Automatica, 6(1975), pp 1-19 ;(h) Global variation criteria for the L_2-stability of nonlinear time varying systems, accepted for publication in SIAM J.Mathematical Analysis ; (i)Converse Schwarz inequality and an instability-related property of feedback systems, accepted for publication in Int.J.System Sci.

V 3. Vidyasagar, M., Some applications of the spectral radius concept to nonlinear feedback system stability, IEEE Trans. Circuit Theory, CT-19(1972), pp 608-615.

V 4. Vinograd, R.E., On a criterion of instability in the sense of Lyapunov of the solutions of a linear system of ordinary differential equations, Dokl. Akad. Nauk SSSR, 84(1952), pp 201-204(Russian).

W 1. Weinberg, L., and Slepian, D., Positive real matrices, J.Math.Mech. 9(1960), pp 71-83.

W 2. Weiss, L., and Infante, E.F., (a) On the stability of systems over a finite time interval, Proc. Nat. Acad. Sci. USA 54(1965), pp 44-48 ; (b) Finite time stability under perturbing forces and on product spaces, IEEE Trans. Automatic Control, AC-12(1967), pp 54-59.

W 3. Willems, J.C., (a) On the asymptotic stability of the null solution of linear differential equations with periodic coefficients, IEEE Trans. Automatic Control, AC-13(1968), pp 65-72 ; (b) Stability, instability, invertibility and causality, SIAM J. Control, 7(1969) pp 645-671 ; (c) The Analysis of Feedback Systems, The MIT Press, Cambridge, Massachusetts, 1971 ; (d) Mechanisms for the stability and instability in feedback systems, Proc.IEEE 64(1976), pp 24-35.

W 4. Willems, J.L., (a) A general stability criterion for nonlinear time varying feedback systems, Int.J.Control 11(1970), pp 625-631 ; (b) Stability criteria for nonstationary feedback systems, Int. J. Control, 7(1968), pp 425-431 , (c) Stability Theory of Dynamic Systems, Nelson, 1970.

W 5. Wintner, A., (a) A criterion for stable characteristic exponents, Quart, Appl. Math. 5(1947), pp 232-236 ; (b) On free vibrations with amplitudinal limits, Quart. Appl. Math. 8(1950), pp 102-104.

W 6. Wong, J.S.W., (a) Some stability conditions for $x'' + a(t)\, x^{2n-1} = 0$, SIAM J.Appl.Math. 15(1967), pp 881-892 ; (b) On two theorems of Waltman, ibid., 14(1966), pp 724-728..

W 7. Wu, M.Y., (a) Some new results in linear time varying systems, IEEE Trans. Automatic Control, AC-20(1975), pp 159-161 ; (b) A note on stability of linear time varying systems, ibid., AC-19(1974), p 162.

W 8. Wu, M.Y., Horowitz, I.M., and Dennison, J.C., On solution, stability and transformation of linear time varying systems, Proc.IEEE 1975 Decision and Control Conference, pp 362-365.

Y 1. Yakubovich, V.A., (a) Solution of certain matrix inequalities occurring in the theory of automatic control, Dokl.Akad.Nauk SSSR 143 (1962), pp 1304-1307 ; Absolute instability of nonlinear control systems II Systems with nonstationary nonlinearities ; the circle criterion, Automation and Remote Control, 32(1971), pp 876-884.

Y 2. Yorke, J.A., A theorem on Liapunov functions using $\ddot{V}$, Math.Syst.

Theory, 4(1970), pp 40-45.

Y 3. Yoshizawa, T., Lyapunov's function and boundedness of solutions, Funkcialaj Ekvacioj 2(1959), pp 95-142.

Z 1. Zadeh, L.A., Time varying networks, I, Proc.IRE 49(1961), pp 1498-1503.

Z 2. Zadeh, L.A., and Desoer, C.A., Linear System Theory, McGraw-Hill, New York 1963.

Z 3. Zames, G., (a) On the stability of nonlinear time varying feedback systems, Proc.NEC 20(1964), pp 725-730 ; (b) On the input-output stability of time varying nonlinear feedback systems, IEEE Trans.on Automatic Control, AC-11(1966), Part I, pp 228-238; Part II, pp 465-476 ; (c) Nonlinear time varying feedback systems - conditions for L_{∞}- boundedness derived using conic operators on exponentially weighted spaces, Proc.Third Allerton Conference, 1965, pp 460-471 ; (d) Stability of systems with sector nonlinearities : A comparison of various inequalities, IEEE Trans.on Automatic Control, AC-13 (1968), pp 709-711 ; (e) Realizability conditions for nonlinear feedback systems, IEEE Trans.on Circuit Theory, CT-11(1964), pp 186-194. (f) Functional analysis applied to nonlinear feedback systems, IEEE Trans.Circuit Theory, 10(1963), pp 392-404.

Z 4. Zames, G., and Falb, P.L., Stability conditions for systems with monotone and slope restricted nonlinearities, SIAM J.Control, 6(1968) pp 89-108.

Z 5. Zames, G., and Kallman, R.R., On spectral mappings, higher order circle criteria, and periodically varying systems, IEEE Trans.on Automatic Control, AC-15(1970) pp 649-652.

Topics in Applied Physics

Founded by **Helmut K. V. Lotsch**

This book series is devoted to research achievements of current interest. Each volume deals with a different topic under the editorship of a recognized authority in the field. It covers application-oriented aspects of the topic under consideration, the basic physical principles being summarized in a comprehensive introduction.
The contributors to each volume are internationally known experts. The publication periods are comparable with those of scientific journals to keep pace with the rapidly accumulating results.

Springer-Verlag
Berlin
Heidelberg
New York

Volume 1
Dye Lasers
Editor: **F.P. Schäfer**
114 figures. XI, 285 pages. 1973
Cloth DM 77,–; US $ 33.90

Volume 2
Laser Spectroscopy
of Atoms and Molecules
Editor: **H. Walther**
137 figures, 22 tables
XVI, 383 pages. 1976
Cloth DM 97,–; US $ 42.70
ISBN 3-540-07324-8

Volume 3
Numerical and Asymptotic Techniques in Electromagnetics
Editor: **R. Mittra**
112 figures. XI, 260 pages. 1975
Cloth DM 72,–; US $ 31.70
ISBN 3-540-07072-9

Volume 4
Interactions on Metal Surfaces
Editor: **R. Gomer**
112 figures. XI, 310 pages. 1975
Cloth DM 78,–; US $ 34.40
ISBN 3-540-07094-X

Volume 5
Mössbauer Spectroscopy
Editor: **U. Gonser**
96 figures. XVIII, 241 pages. 1975
Cloth DM 70,–; US $ 30.80
ISBN 3-540-07120-2

Volume 6
Picture Processing and Digital Filtering
Editor: **T.S. Huang**
113 figures. XIII, 289 pages. 1975
Cloth DM 79,80; US $ 35.20
ISBN 3-540-07202-0

Volume 7
Integrated Optics
Editor: **T. Tamir**
99 figures. XIII, 315 pages. 1975
Cloth DM 79,80; US $ 35.20
ISBN 3-540-07297-7

Volume 8
Light Scattering in Solids
Editor: **M. Cardona**
111 figures, 3 tables
XIII. 339 pages. 1975
Cloth DM 92,60; US $ 40.80
ISBN 3-540-07354-X

Volume 9
Laser Speckle and Related Phenomena
Editor: **J.C. Dainty**
133 figures. XIII, 286 pages. 1975
Cloth DM 94,80; US $ 41.80
ISBN 3-540-07498-8

Volume 10
Transient Electromagnetic Fields
Editor: **L.B. Felsen**
111 figures. XIII, 274 pages. 1976
Cloth DM 92.60; US $ 40.80
ISBN 3-540-07553-4

Volume 11
Digital Picture Analysis
Editor: **A. Rosenfeld**
114 figures. 47 tables.
XIII, 351 pages. 1976
Cloth DM 72,–; US $ 31.70
ISBN 3-540-07579-8

Volume 12
Turbulence
Editor: **P. Bradshaw**
47 figures, XI, 335 pages. 1976
Cloth DM 97,–; US $ 42.70
ISBN 3-540-07705-7

Volume 13
High-Resolution Laser Spectroscopy
Editor: **K. Shimoda**
132 figures. XIII, 378 pages. 1976
Cloth DM 97,–; US $ 42.70
ISBN 3-540-07719-7

Volume 14
Laser Monitoring of the Atmosphere
Editor: **E.D. Hinkley**
84 figures. XV, 380 pages. 1976
Cloth DM 97,–; US $ 42.70
ISBN 3-540-07743-X

Volume 15
Radiationless Processes in Molecules and Condenses Phases
Editor: **F.K. Fong**
67 figures XIII, 360 pages. 1976
Cloth DM 97,–; US $ 42.70
ISBN 3-540-07830-4

Volume 16
Nonlinear Infrared Generation
Editor: **Y.-R. Shen**
134 figures. XI, 279 pages. 1977
Cloth DM 88,–; US $ 38.80
ISBN 3-540-07945-9

Prices are subject to change without notice

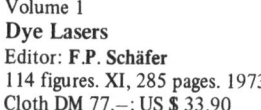

Selected Issues from

Lecture Notes in Mathematics

Lecture Notes in Physics

Vol. 44: R. A. Breuer, Gravitational Perturbation Theory and Synchrotron Radiation. VI, 196 pages. 1975.

Vol. 45: Dynamical Concepts on Scaling Violation and the New Resonances in e^+e^- Annihilation. Edited by B. Humpert. VII, 248 pages. 1976.

Vol. 46: E. J. Flaherty, Hermitian and Kählerian Geometry in Relativity. VIII, 365 pages. 1976.

Vol. 47: Padé Approximants Method and Its Applications to Mechanics. Edited by H. Cabannes. XV, 267 pages. 1976.

Vol. 48: Interplanetary Dust and Zodiacal Light. Proceedings 1975. Edited by H. Elsässer and H. Fechtig. XII, 496 pages. 1976.

Vol. 49: W. G. Harter and C. W. Patterson, A Unitary Calculus for Electronic Orbitals. XII, 144 pages. 1976.

Vol. 50: Group Theoretical Methods in Physics. 4th International Colloquium. Nijmegen 1975. Edited by A. Janner, T. Janssen, and M. Boon. XIII, 629 pages. 1976.

Vol. 51: W. Nörenberg und H. A. Weidenmüller. Introduction to the Theory of Heavy-Ion Collisions. IX, 273 pages. 1976.

Vol. 52: M. Mladjenović, Development of Magnetic β-Ray Spectroscopy. X, 282 pages. 1976.

Vol. 53: D. J. Simms and N. M. J. Woodhouse, Lectures on Geometric Quantization. V, 166 pages. 1976.

Vol. 54: Critical Phenomena. Sitges International School on Statistical Mechanics, June 1976. Edited by J. Brey and R. B. Jones. XI, 383 pages. 1976.

Vol. 55: Nuclear Optical Model Potential. Proceedings 1976. Edited by S. Boffi and G. Passatore. VI, 221 pages. 1976.

Vol. 56: Current Induced Reactions. International Summer Institute, Hamburg 1975. Edited by J. G. Körner, G. Kramer, and D. Schildknecht. V, 553 pages. 1976.

Vol. 57: Physics of Highly Excited States in Solids. Proceedings 1975. Edited by M. Ueta and Y. Nishina. IX, 391 pages. 1976.

Vol. 58: Computing Methods in Applied Sciences. Proceedings 1975. Edited by R. Glowinski and J. L. Lions. VIII, 593 pages. 1976.

Vol. 59: Proceedings of the Fifth International Conference on Numerical Methods in Fluid Dynamics. 1976. Edited by A. I. van de Vooren and P. J. Zandbergen. VII, 459 pages. 1976.

Vol. 60: C. Gruber, A. Hintermann, and D. Merlini, Group Analysis of Classical Lattice Systems. XIV, 326 pages. 1977.

Vol. 61: International School on Electro and Photonuclear Reactions I. Edited by C. Schaerf. VIII, 650 pages. 1977.

Vol. 62: International School on Electro and Photonuclear Reactions II. Edited by C. Schaerf. VIII, 301 pages. 1977.

Vol. 63: V. K. Dobrev et al., Harmonic Analysis on the n-Dimensional Lorentz Group and Its Application to Conformal Quantum Field Theory. X, 280 pages. 1977.

Vol. 64: Waves on Water of Vabriable Depth. Edited by D. G. Provis and R. Radok. 231 pages. 1977.

Vol. 65: Organic Conductors and Semiconductors. Proceedings 1976. Edited by L. Pál, G. Grüner, A. Jánossy and J. Sólyom. 654 pages. 1977.

Vol. 66: A. H. Völkel, Fields, Particles and Currents. VI, 354 pages. 1977.

Vol. 67: W. Drechsler and M. E. Mayer, Fiber Bundle Techniques in Gauge Theories. X, 248 pages. 1977.

Vol. 68: Y. V. Venkatesh, Energy Methods in Time-Varying System Stability and Instability Analyses. XII, 256 pages. 1977.